Computer Programming
for Beginners

Computer Programming
for Beginners
A Step-by-Step Guide

Murali Chemuturi

CRC Press
Taylor & Francis Group
Boca Raton London New York

CRC Press is an imprint of the
Taylor & Francis Group, an **informa** business

CRC Press
Taylor & Francis Group
6000 Broken Sound Parkway NW, Suite 300
Boca Raton, FL 33487-2742

© 2019 by Taylor & Francis Group, LLC
CRC Press is an imprint of Taylor & Francis Group, an Informa business

International Standard Book Number-13: 978-1-138-32048-2 (Paperback)
International Standard Book Number-13: 978-1-138-48096-4 (Hardback)

Library of Congress Cataloging-in-Publication Data

Names: Chemuturi, Murali, 1950- author.
Title: Computer programming for beginners : a step-by-step guide / Murali Chemuturi.
Description: Boca Raton : Taylor & Francis, 2019. | Includes index.
Identifiers: LCCN 2018028243 | ISBN 9781138480964 (hardback : alk. paper) |
ISBN 9781138320482 (paperback : alk. paper) | ISBN 9780429453250 (ebook)
Subjects: LCSH: Computer programming.
Classification: LCC QA76.6 .C436 2019 | DDC 005.1--dc23
LC record available at https://lccn.loc.gov/2018028243

Visit the Taylor & Francis Web site at
http://www.taylorandfrancis.com

and the CRC Press Web site at
http://www.crcpress.com

Contents

Foreword by an Academician

This book is aptly christened as *Programming for Beginners: A Step-by-Step Guide*. The author, Murali Chemuturi, has truly "aimed at making you a star programmer, a professional programmer ..." The perspective of the author has been to hand-hold a person desirous of entering this profession or consolidate upon previously acquired skills in order to carve out and tailor a programmer to meet the requirements of the present-day IT and ICT industry.

Introducing the basics of computers, data and data types, their storage, and retrieval including DBMS as well as computer programs/programming, Murali has elaborated the execution of the program by the computer. Very few authors would dare to talk about the operating systems as well as their operations in a book on Computer Programming. In the next few chapters, the audience is subjected to the rigors of algorithms and flowcharts and handling data in real-life programs while emphasizing and discussing their standard techniques threadbare. This is followed by a detailed discussion of expressions, control statements, I/O, and other related statements explaining the concepts of using advanced aspects of programming.

Murali points out that "The actual implementation of these aspects differs from OS to OS and one programming language to another and all programming languages have not implemented all these aspects." Hence the need to use these concepts in programming necessarily requires that the programmer must *thoroughly* read the relevant parts of the programming language manual, understand it fully, and experiment before implementation.

The subsequent chapters discuss methods of error handling with facilities provided in the OS, interprogram communication coding, debugging, and performance tuning subroutines with best practices and pitfalls as well as building and using libraries and programming device drivers.

Murali has dwelt upon programming multilanguage software and making the software amenable for use in multiple languages which would help ease of programming in the various languages in the world.

Finally, Murali Chemuturi has used all his experience to educate the not-so-knowledgeable entrants to the evolution of programming languages right from the days of COBOL and FORTRAN to the present-day **4GLs (4th Generation Languages)**. Programming standards and guidelines, as well as the scope of these guidelines with specific coding guidelines for each of the programming languages, have also been included for ease of understanding and maintenance. Personal Software Process for monitoring productivity as a metric for quality and efficiency forms the ultimate chapter and includes peer review, coding, and testing methodology.

This book is strongly recommended for the stalwarts in coding as a short lifetime revision for a longer inning as a professional programmer and to the beginners to become "star programmers."

Dr. Suresh Chandra Gupta
University of Mumbai

Preface

Why is a book on the fundamentals of computer programming necessary now, nearly 60 years after serious programming began in the software development industry?

I asked myself this question before I began working on this book. I had already authored books on software estimation, software quality assurance, software design, software project management, requirements engineering, and IT project management, and they are being used as textbooks or suggested/prescribed reading at over 40 universities in 17 countries. I had conducted quite a few corporate training programs, besides lecturing to undergraduate and graduate students in academic institutions. I was a member of interview committees to recruit trainees as well as experienced software engineers. What I had observed is a scenario in which programmers do not have a holistic and comprehensive understanding of software programming. As soon as students join a software development organization upon completing the course as graduate or post-graduate software engineers, they are at sea when beginning their first program! Some professional organizations do not assign project work to the new recruits fresh out of college until they are put through a rigorous internal training program, followed by a test—passage of which is mandatory. The pass mark is usually pegged at 65%! I had professionally conducted such training programs for a few organizations when they recruited graduates, fresh out of college, as trainees.

Dennis Ritchie authored his book on the C programming language, which began with the C language statement "Printf ("Hello World\n")," and it has become a standard practice to begin teaching computer programming with that statement. Some research on teaching methods concluded that the way a child learns his first words is the best way to teach new things to students. The child learns by imitating parents (or those around him) and catching words from them. I agree with this research totally! But I am dismayed at the implementation of that concept, especially for the advanced concepts. While the child initially learns by imitation, the responsible parents put the child through structured instruction the moment the child is able to understand sentences! The child usually goes to preparatory school from the age of 3 years and regular school from the age of 5 years. The structured instruction provides complete information to the student as time passes. I am sorry to say that most of the students out of college presently are like those children who did not go through rigorous instruction after their initial learning.

Unfortunately, our colleges do not take the time to give comprehensive knowledge of programming to students, and the students are, as grown-ups, expected to fill in the gaps through self-study. Most students do not learn anything curricular on their own unless there is going to be a test on that topic!

One more aspect to consider is that the universities do not teach computer programming on a standalone course but as a subject in the computer science program. But when the students come out of college, the bulk of the work they carry out is programming!

All in all, I find that there is a gap in the knowledge of students coming out of colleges teaching computer programming.

Another vital aspect of programming life is its fast obsolescence. A new version of the existing language or a new language appears on the scene every 3 years. If we take an average work-life span of 40 years, the programmer has to learn a new language or a newer version of the language 13 times! The lackadaisical teaching makes it very difficult for programmers to quickly master a new language. The result is that the programmer looks for another trade after two or three new languages.

The learning of newer languages becomes extremely difficult because the initial learning was defective. If we give the comprehensive concepts about programming, rather than giving a few keywords, then learning a new programming language will be easy because it is applying the same concepts with new keywords. If I can, I wish to prevent programmers from falling by the wayside because they are not able to learn a new language quick enough to retain their job. This book is an attempt by me to do just that. This is what this book strives to achieve.

I did quite a bit of teaching and programming myself, and all the books I could find on computer programming were tied to some language or the other. The titles were like *Programming with C, Programming with Visual Basic, Java Programming*, and so on. Most of those books are devoted to getting the reader off the ground to write a simple program and execute it successfully. The concepts and the constructs are not completely covered. I see this as a gap. I want to bridge this vital gap. This book is not tightly coupled with any programming language and, rather, it is not aimed at getting you to write a simple program and execute it to make you feel good about your programming skills. This book is aimed at making you a star programmer, a professional programmer, a Philippe Kahn and a Bill Gates! Perhaps it is an exaggeration, but my intention was to write a comprehensive book to give you a complete idea of all the concepts and constructs of computer programming. Once you finish and master these concepts, I am sure you will be ready to conquer any new programming language that may come on the horizon. I wish you all the best.

Feel free to email me at murali@chemuturi.com and I promise to reply.

Acknowledgments

When I look back, I find that there are so many people to whom I should be grateful. Be it because of their commissions or omissions, they made me a stronger and better person, and both directly and indirectly helped to make this book possible. It would be difficult to acknowledge everyone's contributions here, so to those whose names may not appear, I wish to thank you all just the same. I will have failed in my duty if I do not explicitly and gratefully acknowledge the following persons:

- My parents, Appa Rao and Vijaya Lakshmi, the reason for my existence—especially my father, a rustic agrarian, who by personal example taught me the virtue of hard work and the value of the aroma of perspiration from the brow
- My family, who stood by me like a rock in times of struggle—especially my wife of 44 years, Udaya Sundari, who gave me the confidence and the belief that I can, and my two sons, Dr. Nagendra and Vijay, who provided me the motive to excel
- My two uncles, Raju and Ramana, who by personal example taught me what integrity and excellence mean

To all of you, I humbly bow my head in respect, and salute you in acknowledgement of your contribution.

Murali Chemuturi

About the Author

Murali Chemuturi is an information technology and software development subject matter expert, hands-on programmer, author, consultant, and trainer. Since 2001, he has been offering consultancy on information technology and training to organizations in India and the USA through Chemuturi Consultants. Chemuturi Consultants also offers a number of products to aid project managers and software development professionals, such as PMPal—a software project management tool—and EstimatorPal, FPAPal, and UCPPal, a set of software estimation tools. Chemuturi Consultants also offers a material requirements planning software product, MRPPal, to assist small- to medium-sized manufacturing organizations to efficiently manage their materials.

Prior to starting his own firm, Murali gained over 15 years of industrial experience in various engineering and manufacturing management positions. He then gained more than 30 years of information technology and software development experience. His most recent position prior to starting his consultancy was Vice President of Software Development at Vistaar e-Business Pvt., Ltd.

Mr. Chemuturi's undergraduate degrees and diploma in Electrical Engineering and a graduate ship in Industrial Engineering, and he holds an MBA and a Post Graduate Diploma in computer methods & programming. He has several years of academic experience teaching a variety of computer and IT courses, such as COBOL, FORTRAN, BASIC, Computer Architecture, and Database Management Systems. He was inducted into the Hall of Fame by the CSI, Mumbai Chapter, in December 2016.

Mr. Chemuturi has authored three books, namely *Software Estimation: Best Practices, Tools and Techniques for Software Project Estimators*, *Mastering Software Quality Assurance: Best Practices, Tools and Techniques for Software Developers*, and *Mastering IT Project Management: Best Practices, Tools and Techniques*, published in the USA by J. Ross Publishing, Inc. He has also authored a book on *Requirements Engineering and Management for Software Development Projects*, published in the USA by Springer Science+Business. He co-authored another book with Thomas M. Cagley, Jr., titled *Mastering Software Project Management: Best Practices, Tools and Techniques*, published by J. Ross Publishing, Inc of the USA. He also authored *Software Design: A Comprehensive Guide to Software Development Projects* published by CRC Press in the USA.

Murali is a senior member of IEEE, a senior member of the Computer Society of India, a Fellow of the Indian Institute of Industrial Engineering, and a well-published author in professional journals.

1

Introduction to Computers

What Is a Computer?

People may say that perhaps 20 years ago we might have needed this chapter, but not now! Perhaps they are right, but in my interactions with people, I have seen misunderstandings or different understandings about computers and what they can do. I have seen computer science graduates look at me incredulously when I asked them this question, but when pressed, they could not provide me with a credible answer. So, by defining and explaining what a computer is, I, the author, and you, the reader, would be on the same page.

One definition is, "one who computes is a computer." Yes, a computer does compute. And some time back, organizations had employees with the designation of "computer." He/she was usually in the finance and accounts department checking all the important computations carried out by other employees in the department. A computer was originally developed to solve mathematical problems, but today's computers do more than just solve mathematical problems. So, this definition, while it is correct, does not adequately define the computers of today.

Another definition of computer is, *"A computer is a data processing tool."*

The key words in this definition are:

1. *Computer*: This describes the hardware part of the computer. It is a fast, electronic machine that is versatile and assists human beings in a variety of applications. It works accurately and diligently. By diligence, I mean not being subject to fatigue or monotony. It performs the computations a million times with the same accuracy and speed as it would the first time.

2. *Data*: Data is facts and figures about entities. An entity is a person, place, thing, or transaction. Data is comprised of attributes describing the entities. A person is described by his/her personal attributes, such as height, weight, educational qualifications, vocation, address, age, place of work, and so on. A place like a building is described by its address, location, purpose of its usage, number of people using it, number of rooms, other facilities, number of floors, and so on. A town is described by its location, county, state, country, size of its population, important landmarks, famous personalities, its history, and so on. A business can be described by its date of incorporation, its location, address, its outputs, its customers, its vendors, and so on. A thing can be an object, a machine, a tool, an appliance, or something like that. It is described by its make, model,

date of manufacture, date of purchase, its function, and so on. A transaction can be any interaction between entities. It can be a purchase transaction, a sales transaction, a payment transaction, a receipt transaction, a taxation transaction, a contract, and so on. All transactions are described by the date, type, entities involved, place of transaction, terms and conditions, and so on. All these individual attributes are data.

3. *Processing*: Processing is the transformation of data into information using mathematics and logic so that it becomes meaningful to the concerned individuals. Processing includes: solving mathematical problems and giving solutions; arranging the data in some desired order; classifying the data into the desired classes; rendering the pictures/graphics as desired; communicating with the world as designed; and reporting the results of the processing as designed.

4. *Tool*: A tool is an aid for an artisan. A tool does nothing on its own. An artisan can do something on his/her own, but a tool enhances the efficiency and the effectiveness of the artisan. Without a pipe wrench, a plumber would not be able to do many things a plumber is asked to do. But the mere possession of a pipe wrench does not make one a plumber! A tool must be in the right hands to be effective. In that sense, the computer, too, must be used by an expert in computers to be able to derive any results. But you may say that the computer is used by people who are not experts in computers. True—the computer is like an automobile. To make use of a car, one need not be an automobile engineer. But the individual needs to know how to drive and be licensed to drive the car. To make use of a personal computer loaded with the required programs like the Microsoft Office suite, one needs to be adept at using that suite of applications.

We can use this definition to understand computers and move forward.

Bits and Bytes

Before we move on to other topics, we need to first understand the terms "bit" and "byte," as we will be encountering these terms frequently. A bit is the short notation of "binary digit." As computers work by using electricity for processing, representing only two digits is easily achieved. Take a switch; its on position represents 1 and its off position represents 0. Using a binary numbering system is facilitated by these two states. While all numbers from our usual numbering system exist (that is, the decimal numbering system in which we have 10 digits beginning with 0 and ending with 9), the computer only uses the binary numbering system that has only two digits. A code, using binary digits, is usually used to represent all characters inside the computer so that the machine can understand and perform the required processing. Now that we understand that "bit" is the short form of "binary digit," and that a binary numbering system is used inside the computer to represent the characters, we can move forward.

The first system of codifying characters using a binary numbering system used four binary digits; this was called BCD (Binary Coded Decimal), and this was used to codify the numbers. It allowed for 16 codes (2^4), but it was insufficient to codify all alphabets in addition to numbers. A set of 8 bits was settled upon by the company IBM to codify all characters for its mainframe computers, and it was called the EBCDIC (Extended Binary Coded Decimal Interchange Code). IBM continues to use this code in its mainframe computers to this day. Then the ASA (American Standards Association, now, American National Standards Institute [ANSI])) developed another system of codifying characters for use in computers that is called ASCII (American Standard Code for Information Interchange). It originally used a 7-bit code that allowed 128 (2^7) characters to be codified. Presently, ASCII code is extended to 8 bits, adding 128 more characters to be codified. ASCII is used in all personal computers and those computers built using VLSI chips for the processor. There can be exceptions, but most of computers, other than mainframe computers, presently use ASCII code.

The set of 8 bits used for codifying characters is called a "byte." A byte can store one character inside the computer. So, a bit is the short form of "binary digit," and byte is a set of 8 bits and can store one character.

Components of a Computer

A computer is made up of two main components, namely, the hardware and the software. Let us first look at hardware and then move on to software.

Hardware—CPU, RAM, Bus, and System Clock

Computer hardware has gone through a very rapid evolution never before witnessed in any other field of human endeavor. Computers came into being slightly before I was born, and we were speaking with awe about computers during my early adulthood. But now, I personally own a couple of laptop computers for my usage alone! The components of a computer of today are very different from those of the earlier computers.

CPU

At the heart of the present-day computers is the transistor, but the transistors are integrated into a chip called the integrated chip. Depending on the number of transistors integrated into the chip, they are called LSIC (Large Scale Integrated Circuits) and VLSIC (Very Large Scale Integrated Circuits). Presently, they achieved the capability to pack more than a million transistors onto a single chip. The base of the chip is basically silicon crystals on which the electronic circuits are engraved and transistors are etched. A VLSIC chip is nothing less than a miracle. Because of this integration, not only did the price of the computer come down, but its size also was brought down significantly. While the first computer occupied a large room, its equivalent or even a more powerful computer is accommodated in a handheld tablet today.

A computer consists of a processor unit, Random Access Memory (RAM), a control unit, input devices, and output devices. Strictly speaking, the processor unit with its RAM is the computer, and the rest are peripheral devices to the computer.

The processor unit is referred to as the Central Processing Unit, or simply, CPU. It is the heart of the computer system. The CPU carries out all the data processing operations. All the data necessary for the CPU to process needs to be stored in the RAM.

The CPU and the RAM are very tightly coupled. The CPU contains what are referred to as "registers." Each register is a block of memory. Some of the important registers of a processor are:

1. *Accumulator*: This register stores the result of an operation.
2. *PC—Program counter*: This register is used to store the address in RAM of the next program instruction to be executed.
3. *Instruction register*: This register is used to store the instruction currently under execution.
4. *Data registers*: These registers can be in multiple numbers each storing the value of one data item.
5. *MAR—Memory address registers*: These registers store the address in RAM for the next data items to be fetched.
6. *General purpose registers*: These registers can be used for any purpose, which include storing the instruction, data values, addresses, or results of operations.
7. *Stacks*: These are a special type of register used specifically in the solution of mathematical expressions.
8. *Pointers*: These are primarily used to store addresses of locations in the RAM.
9. *Flag*: It stores only one of the two possible values: True or False.
10. *Stack pointer*: When the stack is not inside the CPU, this register stores the address in the RAM where the stack is located.

The CPU performs all the processing by moving data and instructions into and out of the registers. How all that movement and processing happens is beyond the scope of this book. I have included this information to give you an insight into the CPU. The size of the register is specified in the number of bits its can store. The size of a register is generally in lots of bytes; that is, a register is either 8, 16, 32, or 64 bits.

Bus

The other component of the processor is the bus. Bus is in fact the short form of bus-bar. "Bus-bar" is a phrase used in electrical engineering. Inside electricity-generating stations and distribution stations, bus-bars are used to allow electricity to flow from one terminal to another. Other electrical equipment can place taps on the bus-bars to draw electric current as required. The electrical bus-bars are insulated or uninsulated heavy metal rods, depending on the current that flows inside them. For a three-wire system, three bus-bars would be needed. That is, the number of bus-bars needed would be determined by the number of parallel circuits required.

The bus in the computer is built on the same premises. A bus is used to move data and instructions to and from the CPU to RAM. Each line in the bus allows transmission of one bit of information. The capacity of the bus is defined in the number of lines in the bus. It is generally expressed as a 16-bit bus, 32-bit bus, or 64-bit bus, and so on.

Now the capacity of a processor in the computer is expressed in terms of the size of the registers inside the CPU and the capacity of the bus. If the size of the register in the CPU is 32 bits and the size of the bus has 16 bits, the processor is referred to as 32/16 bit. Usually, the sizes of the registers and the bus are maintained at the same level, but difference in their sizes is also possible.

RAM

Then the other important component of the processor is the primary memory, or RAM. While RAM is physically outside the processor, it is considered part of the CPU. Magnetic cores of annular-ring shape were used for RAM in earlier computers but are now replaced by silicon chips, commonly referred to as the memory chips. Each element of RAM has only two states, which is a true or false state; in other words, it stores a value of zero or one. Computers in the earlier days had very limited RAM, but now computers have a large amount of RAM. Having a large amount of RAM decreases the response time of the computer and increases the overall output.

The capacity of the RAM is expressed in terms of bytes: kilobytes (KB), megabytes (MB), and gigabytes (GB). Presently, even tablets have gigabytes of RAM. A kilobyte has 1024 (2^{10}) bytes, a megabyte has 1,048,576 (2^{20}) bytes, and a gigabyte has 1,073,741,824 (2^{30}) bytes. In common parlance, a KB is taken as a thousand bytes, an MB is taken as a million bytes, and a GB is taken as a billion bytes.

How much RAM do we need if we want to have a computer that does not make us wait endlessly for a response? It should be more than the sum of RAM required for the operating system and all the programs we wish to run concurrently. If you are using a windows-based PC or laptop, these details are given in the Task Manager, which you can see by right-clicking the task bar and selecting Task Manager.

System Clock

System clock is another component of the CPU. The system clock releases a pulse at regular intervals of time. The CPU performs one operation with every pulse released by the system clock. Therefore, the number of operations performed by the CPU depends on the number of pulses released by the system clock. The speed of the system clock is expressed in hertz; kilohertz (kHz), megahertz (mHz), and gigahertz (gHz). All present-day processors have speeds greater than a gHz. A hertz is the unit of measure for frequency. Frequency is the number of cycles taking place per unit time. In the present case, a hertz is equal to one cycle per second. A system clock speed of 1.6 gHz means the clock releases 1.6 billion pulses per second and the CPU performs 1.6 billion operations per second.

Input/Output Devices

It is the I/O (Input/Output) devices through which we interact with the computer. The keyboard, mouse, and display screen are most commonly used I/O devices and are familiar to any computer user. We can safely say that a computer is useless without the I/O devices. The development of versatile I/O devices is the reason for the popularity of computers. There are many other I/O devices for a computer and some of them are enumerated here:

1. *Keyboard and mouse*: I am sure that everyone is familiar with these two devices that provide input to the computers.
2. *Display screen*: I am once again sure that everyone is familiar with this output device. There are, of course, a variety of display screens, including cathode ray tubes, which are on the way out, LCD (Liquid Crystal Display) screens, plasma screens, and LED (Light Emitting Diode) screens. Earlier the screens were able to display only character data in one color. The current display screens display stunning colors as well as graphic data.

3. *Printers*: There a variety of printers available on the market for different applications. Heavy-duty printers are used for bulk printing needs. These include line printers or line matrix printers. Inkjet printers and laser printers are used for low-volume personal printing purposes. Photo printers are used for printing photo-quality outputs either on paper or polyester films. Photo printers and inkjet printers can print in color. Usually, the heavy-duty printers print only in black and white. Printers are only output devices.

4. *Magnetic tapes*: Once upon a time, magnetic tapes (shortened as "mag tapes") were the primary storage device driving the computers. In those days, in movies, computers meant whirring magnetic tape drives. The tape drives have gone out of use more or less, except as backup devices. Presently, data cartridges, which are basically magnetic tapes of a much smaller size, having capacity to store a hundred GB of data are used to take backups of corporate data.

5. *Plotters*: Plotters are used to generate engineering drawings. They come in five sizes, namely, A4 (8.3″ × 11.7″ − 0.0625 sq m), A3 (11.7″ × 16.5″ − 0.125 sq m), A2 (16.5″ × 23.4″ − 0.25 sq m), A1 (23.4″ × 33.1″ − 0.5 sq m), and A0 (33.1″ × 46.8″ − 1 sq m). These are standard sizes for engineering drawings. A larger size plotter would be able to produce its size and all the lesser sizes of drawings. Thus, an A0 plotter can be used to produce not only A0 size drawings, it can also produce A1, A2, A3, and A4 size drawings. Plotters usually come with color capability, but they would not be able to produce photo-quality outputs. These are basically used for producing engineering drawings.

6. *Computer networks*: As computer networks have come of age and a large part of the world is now connected, most of the computers access Internet at a minimum. Computers output information onto the network and receive input from the network. Networks are both input and output devices.

7. *Machines*: Now, a variety of machines are controlled by computers. Of course, the computers used for machine control are slightly different from commercial computers in their ability to withstand extremes of climate and the strict response times needed by the machines. Today, not just airplanes, rockets, and machining centers in factories, but also mundane devices including cars, refrigerators, and washing machines are controlled by computers. Machines are mostly output devices except sending responses about the successful or unsuccessful execution of computer instructions and error conditions.

8. *CD/DVD drives*: These are used for storing information for future use. These act as both input as well as output devices.

9. In the past, we were using punched card machines for input and output from computers, floppy drives and floppy disks for backup storage, and teleprinters for both input and output. Paper tapes and paper tape readers were also used for driving NC (numerically controlled) machines in factories. All these devices are more or less obsolete now. They could be in use in some remote places, but not in large numbers.

10. *Robots*: Robots are versatile automatic machines that can be programmed for a variety of tasks. These are used in factories for manufacturing. Robots can be solely output devices receiving computer instructions to perform assigned operations, or they can be solely input devices providing input like sounds and pictures to the computers. The Mars rovers that are collecting enormous data from the planet Mars are, in fact, robots of a special kind. They even analyze Martian soil!

Computers are also used in process control in factories, such as petroleum refining, fertilizer production, chemical manufacturing, drug/pharmaceutical production, and so on. In fact, computers are used in any automation where the parameters are programmable using numbers. The I/O devices listed earlier are by no means the comprehensive listing. There are many devices that can be interfaced with computers. In fact, machine designers are designing in such a way that many machines can be interfaced with computers so that the processing capabilities of computers can be profitably utilized.

Software

While hardware is the visible component of the computer, software is the invisible component. Software is the set of computer programs that run the computer system and process data. Hardware performs the operations, and software directs the hardware operations. While hardware components are like the individual musicians in a concert, the software is akin to the conductor. The melody is achieved by the conductor and the musicians obeying the conductor. The software has primarily three components, namely, the firmware, the operating system, and the application software.

Firmware

Firmware is the basic software that is usually stored on a ROM (Read Only Memory) chip. This software is generally supplied by the hardware manufacturer along with the computer. As soon as the computer is switched on, this software takes over and performs the following functions:

1. It stores the information about the configuration of the system and applies it at the time of switching on.
2. It checks the configuration of the system to ensure that it is not changed in an unauthorized manner since the last time it was checked.
3. In some cases, it also stores the hardware password, then prompts for entering the password, ensures that the entered password is correct, and allows access to the rest of the operations only upon entering the right password.
4. It checks all the hardware components to ensure that all are working as they should be.
5. When anything is found to be out of place, it displays error messages, allows for correction of the defects, and prevents switching on a faulty computer system.
6. It then loads the computer operating system and passes on the control to it for further operations.

Operating System

The second component of the software is the OS or the Operating System of the computer. The OS controls and governs the hardware resources of the computer system. The OS interfaces between the users and the hardware resources of the computer system. An OS consists of four components apart from the user interface:

1. Processor management
2. Memory management

3. Device management

4. Information management

The OS comes in two flavors, namely the single-tasking OS and the multi-tasking OS. In a single-tasking OS, the computer system performs only one process at a time. The old MS-DOS (Microsoft Disk Operating System) and its predecessor, CP/M (Control Program for Microprocessors), were both single-tasking OS. Present-day Windows OS is a multi-tasking OS.

Multi-tasking OS comes in two varieties, namely, the single-user OS and multi-user OS. In single-user OS, only one user can use the computer at any given time. Windows OS is a single-user OS. It permits only one user to use the computer at any given time. Two or more users cannot use it concurrently. Multi-user OS permits multiple users to use the computer system concurrently. UNIX and its variants, as well as mainframe computer OS, are multi-user OS.

Processor Management

The processor management module of the OS manages the processor of the computer system. The processor executes processes. A process is a program in execution. In all multi-tasking OS, there are multiple processes waiting for the processor to execute them. The functions of the processor management module are:

1. Maintain a queue of all the waiting processes.
2. Allocate a processor to each of the waiting processes for a specified amount of time and then deallocate the process.
3. Using a fixed time slice for each of the processes, ensure that each of the waiting processes is allocated with the processor.
4. Keep track of the stage of execution for each process and ensure that each waiting process is in the queue until each process completes its execution.
5. Enforce the policy of processor allocation to the processes scrupulously.
6. Monitor if a process is hogging the CPU time and raise an alert so that it can be corrected.
7. Monitor if a process is becoming tardy, or being pushed to the end of the queue for a long time, and raise an alert so that corrective action can be taken.
8. Monitor the load on the processor and raise alerts when needed so that corrective action can be taken. When too many processor-intensive processes are waiting for execution, no process may get completed and it may look as if the system is frozen. With a good processor management module, the system administrator would get an alert to correct the situation.
9. Ensure that the processor is utilized efficiently.
10. In multi-processor computer systems, ensure that all the processors present in the computer system are uniformly loaded and that no single processor is either overloaded or under-loaded.

In multi-tasking OS, multiple processes are waiting for the time of the processor. The time of the processor is shared between the processes. That is why those OS are called as time-sharing systems. The method of allocating processor to processes varies from OS to OS.

In most business computer systems, a round-robin method is used. In this system, each process gets the processor for a fixed amount of time before it is moved to the end of the queue. The process will get the processor once again after all the waiting processes in the queue ahead of it get their time slice with the processor. So, a waiting process is allocated to the processor a number of times before its execution is completed. A slight variant of the round-robin method is to assign priorities to each of the processes. A process with a higher priority gets more time slices per round than the processes with a lesser priority. Generally, OS processes would have a higher priority than processes of application software. Some OS allow the system administrator to manually assign a higher priority to a process. In some OS where the response time is specified, as in the case of machine-control computers, the processor allocation method would be based solely on priority.

Memory Management

In SPA (Stored Program Architecture) computers of the present day, the program under execution needs reside in the primary memory. A multi-tasking OS has multiple programs in its memory. Each of the programs will have these components in the RAM until the execution of the program is completed:

1. The executable code of the program in its entirety.
2. One set of all data elements for all the variables declared inside the program.
3. For each of the data files used in the program:
   a. The address and other details of the data file on the disk.
   b. One record space for each type of record in the data file.
   c. The disk address of the next record to be read.
4. For each database used in the program:
   a. The details of the database.
   b. One record space for each of the tables opened.
   c. The address of the next record to be read from each of the tables that were opened.
5. If any machine or port is used by the program, its details also need to be in the RAM.

RAM has to be allocated for these spaces for each program under execution along with the space required for the OS. It is often the case that the available RAM is insufficient to hold the details of all programs that are under execution. In such cases, this module swaps some programs on to the disk to a specified area called the "swap" area and retrieves it whenever it is once again required by the processor. Usually, the programs requiring an I/O operation are swapped. For this purpose, RAM is organized into segments. Each segment is divided into pages or blocks. Usually a page or block of RAM is 1 KB, but it can vary from computer to computer. In modern computers, demand paged memory management is used. In this method, the pages that are not demanded by the processor are swapped to the disk and the pages demanded by the processor are brought from the disk to the RAM. When the RAM is too little, the swapping in and out becomes excessively high and the OS would only be doing the swapping activity. This is referred to as "thrashing." The system, as a result of thrashing, becomes too slow.

The memory management module of the OS performs the following functions:

1. Monitor the RAM available, ensure that all bits are in working condition, and raise alerts whenever any part of the RAM is not working.
2. Maintain a table of all the pages of memory available and their allocation details.
3. Allocate RAM to programs in execution as required and available.
4. Manage the process of swapping in and out smoothly and automatically.
5. Ensure that the RAM allocated to a program is not violated by any other program and that the integrity of the program data is not violated.

Device Management

This module is also referred to as the I/O (Input/Output) management, as every device connected to a computer is either an input device or an output device. Some devices, like the disk, are both input and output devices. Usually, each device connected to a computer system is controlled by its control circuit, which is connected to the system bus. Each device comes with its own device driver software. The OS of the computer would have sockets to plug in the device driver software and whenever a request is placed for the use of the device by a program, or the user of the device itself, the OS hands over the request to the concerned device driver software and monitors the device.

The device management module of the OS would perform the following functions:

1. Provide sockets on the OS to enable plugging in the device driver software for any compatible device.
2. Maintain a list of devices connected to the computer system and monitor their functioning.
3. Communicate with the device, including:
   a. Receive communication from the device.
   b. Decipher the device from which the communication is received.
   c. Take the action requested by the device.
   d. Communicate the result back to the device.
   e. Raise alerts whenever the device is malfunctioning or not functioning.
4. Interface the processes with the devices:
   a. Receive a request for the device from the process.
   b. Decipher to which device the request should be routed.
   c. Transmit the information to the concerned device.
   d. Receive the response from the device and communicate it back to the process.
   e. Raise alerts if the device is either not functioning, malfunctioning, or busy servicing a request from another process.
5. Maintain a buffer memory to bridge the gap in speeds of the computer and the device.
6. Manage the communication protocol between the computer and the device.

Information Management

In the present day, magnetic disks are used as the main secondary memory or secondary storage of the computer information. RAM inside the computer is volatile and all its information is lost when the power is switched off. Therefore, the data is stored on a medium that can retain information even when the power is switched off. CDs (Compact Disks), DVDs (Digital Video Disks), flash memory (also referred to as solid state disk/memory), and magnetic disks are used for this purpose. The disks are organized into tracks, and each track is divided into sectors or blocks.

When the disk is formatted, the computer performs the following functions:

1. Scan the entire disk for errors.
2. Assign a unique number to each of the blocks of information.
3. Create a VTOC (Volume Table of Contents) on the disk. VTOC is also called FAT (File Allocation Table), File System, and so on. This table would contain the information of the file to which space is allocated, the details of the blocks of the disk allocated to the file, and other relevant information.
4. Create a table to hold the information of damaged areas (called the bad blocks) of the disk and enters the details of all damaged areas of the disk into this table. This table would be used to ensure that these bad blocks are not allocated to any file.
5. Sometimes, disk partitions can be created on a single volume of disk. When the disk is partitioned into multiple volumes, a partition table is maintained and is filled with the details of each partition, with the beginning address and the ending address of the blocks allocated for each of the partitions.
6. Sometimes, the disk is made bootable; that is, the firmware uses this disk to load the operating system into the RAM. In such cases, the formatting of the disk loads the system boot information into the first block of the disk and the number of blocks required contiguously.

Now the disk is ready for use inside the computer. Now the information management module of the OS uses this information to load information into this disk. The information management module of the OS handles the following functions:

1. Allocate disk space to the files efficiently.
2. Allocate additional space as required for the expanding data files while maintaining the linkage between the existing data and the new data in such a way that retrieval is smooth without requiring human intervention.
3. Deallocate space when data files shrink in size and add the released blocks of space to the list of free space available on the disk so that they can be allocated against new requests for space.
4. Utilize an efficient allocation algorithm so that the fragmentation of files is minimized.
5. Handle the operations of disk write and disk read efficiently with the least possible access times and maximizing the data transfer rates.
6. Scan the disk as scheduled and locate bad blocks, if any, and add them to the bad blocks table to prevent storing information on them.

Network Management

With the coming of age of the computer networks and the Internet, each computer OS is equipped with a module for managing the network. A server that controls the network would have network management software as a large and special module. Other computers which are not servers would have a module for managing network communications. It would perform the following functions:

1. Receive communications from the network.
2. Inspect each packet of information received over the network and decide if the communication is intended for this computer.
3. Accept all packets of information addressed to this computer and receive them and assemble them into a file and store it at a desired place.
4. Reject all the packets of information not addressed to this computer.
5. Prepare information to be transmitted from the computer onto the network into packets complying with the network protocol.
6. Transmit the packets of information onto the network and receive acknowledgement that the packets are received without defects.
7. Ensure that the data transmission from and to the computer are completed successfully without errors.
8. Carry out housekeeping tasks associated with the data transmission on the network.

Miscellaneous Utilities

Utilities are miscellaneous programs usually bundled along with the OS by the supplier. These include file copy, file delete, list the contents of the disk, display the contents of a data file, back up data files, restore data files, system administration utilities, and so on. Each supplier of OS supplies a different set of utilities needed for the OS. The purpose of these utilities is to make it easier for the users to use the OS efficiently.

Application Software

The software that is neither firmware nor the OS is application software. It works above the OS, utilizing the services provided by the OS. The office suites, database management systems, media players, and all such packaged software supplied on a CD or DVD also fall under application software. Any software developed specifically for our purpose is also application software. Ticket reservation software, gaming software, supply chain management software, customer relationship management software, accounting software, and all other software that we use comes under application software.

While computer hardware, firmware, and OS facilitate use of computers, application software is the one that carries out real work that results either in revenue or in reduced costs for the organizations. Most of the software development work that is carried out in the world is for developing application software.

What Can Computers Do?

Computers can accomplish only one function and that is to process data. In fact, it is the I/O devices that use the processed data and perform all the functions attributed to computers. Basically, computers carry out these two functions:

1. *Number crunching*: All mathematical computations can be carried out by computer. But they need programming. We have some software packages like the SPSS (Statistics Program for Social Sciences), R software, and spreadsheets like the Excel that have programmed many mathematical formulae and allow us to use them to perform our mathematical analysis without programming the computer.

2. *Programmed decision making*: Computer can compare two values, either numerical or character, and determine if they are equal or not. When they are unequal, it can also decide which one has a higher/lower value. Using this feature, we can program decision-making logic into our computers and utilize the computer in decision making. One example for programmed decision making is the evaluation of the examination answer scripts and declaration of results.

Using these two functions, and the capability of the I/O devices, we achieve a variety of functionalities. We use the graphics hardware and display screen in engineering designs to produce drawings; we use the number crunching capability to evaluate engineering designs; and with number crunching capability and programmed decision-making capability, we use computers in airplanes and rockets, too.

Where is the intelligence that computers are much touted for? It is there in the computer programs and the storage media as data. Programs make use of the data, the standard data in the secondary media and the input data coming from the input devices. Computers crunch the numbers, make comparisons complying with the rules in the programs, and give out intelligent decisions.

Computers are diligent and are free from monotony and fatigue that inhibit human beings from doing certain functions. Crunching large numbers using complex formulas can be performed by human beings, but human beings, being prone to error, are likely to commit mistakes. Once programmed correctly and a sample is checked, computers can perform the operation without any mistakes, even after a million times, without suffering from monotony and fatigue. Again, the human beings, not considering the exceptions, are much slower than computers, especially in performing a large number of computations. Thus, the computers have come to perform many operations that were not thought of initially by the people who built the first set of computers.

Thus, the computers can crunch numbers and make programmed decisions diligently and very quickly. As programmers, we need to be aware of this fact.

How Do Computers Work?

Present-day computers are referred to as the digital computers. That means they function by counting. Counting is basically incrementing a number by a specified value. Counting includes counting backwards too. Suppose we'd like to add 5 to 4. What it tells the computer is to increment the number 4 five times with an increment value of 1. Supposing you wished to subtract 3 from 5, the computer would decrement 5 three times with a decrement value of 1. If you wished to multiply 3 by 4, the computer would perform it in three installments; in each installment, it increments the number three times. Initially, it starts with 3 and increments it three times, obtaining 6 as the result. In the second installment, it takes 6 and increments it 3 times getting 9. In the final installment, it begins with 9 and increments it 3 times getting the result of 12. Why did it select 3 installments? It selected 3 installments because, 3×4 is the same as $(3 + 3 + 3 + 3)$, and the first 3 is already available. So, it needs to add 3 only three times. Division is just the reverse of the multiplication process.

Extending this procedure, we can solve all mathematical equations. When you attempt to solve a long, complex mathematical equations in this manner, it would require a significantly large number of increment/decrement operations. But, as the computer is capable of performing millions of operations per second diligently, we get the answers very quickly.

Basically, computer is simple at heart but is very diligent, shorn of fatigue and monotony. So, we are able to use it for a variety of purposes to solve our problems using computers.

Who would tell the computer to perform all these operations? The computer programs pass on the instructions to the computer. Some procedures are built into the OS; some are built into the compilers; some are supplied as libraries; and some need to be programmed by the developers of application software.

Final Words

In these days, there is no one in the developed countries that has not heard of computers. But most would understand the computer the way a car or an airplane or a ship is understood. We could have seen one, used one, and suffered with one, but most of us do not know how a car functions or an airplane flies or a ship floats, let alone their internal components and functioning. Aeronautical engineering, mechanical engineering, marine engineering, or electrical engineering fields are not amenable for accountants, lawyers, or computer programmers to undertake work, but computer programming is such a field in which people who did not have a formal degree in computer science are working in it. You can find people from all fields working as computer programmers. Why? Because it is not necessary to know the internal workings of a computer to be able to write computer programs. It is adequate if you are expert in a programming language and can understand the user requirements. That is the reason for me to include a rudimentary chapter on an introduction to computers. Well, this information is by no means comprehensive. Perhaps I abridged about 1500 pages of information into the pages of this short introduction to arouse your interest so you would read it and become proficient in computers some time later. I believe that it is essential to understand computers to be a good computer programmer.

2

Introduction to Data and Data Types

What Is Data?

Data, as we defined in Chapter 1, is facts about entities. An entity can be:

1. A person
2. A location
3. An object
4. A system
5. A piece of equipment
6. An item of material
7. A transaction
8. A town
9. It can be anything that has some attributes by which it is described!

Facts can be in figures or be descriptive. For example, a person has a name that is descriptive in nature; has a weight that is expressed in numbers; has an educational qualification that is descriptive; has an income that is described in figures; has a date of birth that is a special type of number; and has a title that is, again, descriptive. In this manner, all entities have some facts associated with them.

While each entity has facts associated with it, we are not interested in all entities. We are interested in those entities whose data needs to be stored in our computer system, processed, and reported as required.

Basic Data Types

Data is basically of two types:

1. The data that human beings use.
2. The data used solely by the computer and its I/O devices.

As programmers, we perhaps need to handle both types of data. Then the data used by human beings is of two types, namely:

1. *Character data*: The data that contains alphabets and perhaps numbers, too.
2. *Numeric data*: The data that is expressed in numbers.

Character Data

Character data includes any character that can be input to the computer. What are characters in the context of computers? It can be anything accepted by the computers. IBM uses EBCDIC (Extended Binary Coded Decimal Interchange Code) codification of characters for its mainframe computers. It is an 8-bit code that allows 256 characters to be codified. Later, the American Standards Association finalized the standard for codification of characters for use in computers and called it ASCII (American Standard Code for Information Interchange). It was originally a 7-bit code but has now extended to 8-bits and allows 256 characters to be codified. It codifies not only alphabets (uppercase and lowercase), numerals (0–9), and other special characters that are both humanly readable (like space, (,), [,], and so on), but also those that are not humanly readable (like the enter button, CTRL, ALT, and so on). Character data can also contain those characters that are not humanly readable in addition to those humanly readable characters.

How do computers treat character data? The characters that are not humanly readable provide special input to computers. They are generally used in combination with other characters to give special commands to computers. Humanly readable characters are used to denote names of entities to be stored, retrieved, and searched in the computers. The treatment of character data depends on the compiler's definition of character data of the specific programming language.

In the earlier days, COBOL defined character data into two classes, namely, alphabetic data and alphanumeric data. Alphabetic data allowed only alphabets and a blank space. Alphanumeric data allowed alphabets, numbers, and other humanly readable characters.

BASIC and C languages allowed for the definition of a single character, which allowed any character to be input to the computers.

Character data is usually defined as a string of words. A word is a contiguous set of characters. A word is separated from another by a delimiter, which usually is a blank space. Character data that is humanly readable is usually stored inside the computer. It is used to search for a specific set of data inside a large group. It is also used to make a logical decision based on the word or a group of words. It can also be added; that is, two words can be added (or concatenated) to result in a single word. It can also be used to pick parts of words and make a new word. The only arithmetic operation possible on character data is addition (or concatenation). Other arithmetic operations including multiplication, division, subtraction, and so on are not possible on character data. True, some languages permit performance of all arithmetic operations on character data, but they do not promise reliable results or any practical purpose. Character data is generally used as:

1. *Strings*: A string can consist of multiple words. When we use strings in programs, we need to enclose the string between quote marks ("...."). Usually the length of a string is restricted to 255 characters or, in other words, one-fourth of a one kilobyte.

2. A memo or long textual matter is used to store documents or long explanations inside a database. It can contain lines in addition to words and strings, numbers, and other humanly readable special characters such as parentheses, full stops, commas, and so on.

3. *Special strings*: Recently, this data type has come into existence. The URLs (Uniform Resource Locators) used for website addresses and email addresses form part of this data type. These are used for navigating to a website or sending an email.

Numeric Data

Numeric data is numbers expressing the value of some attribute of an entity. As you perhaps know, there are real numbers and imaginary numbers. A computer can handle only real numbers on its own. What is an imaginary number? The square root of any negative number is imaginary. However, if you can come up with a procedure for handling imaginary numbers, then, perhaps, it can handle them. If during the execution of a program, the computer comes across an imaginary number, an error is thrown up.

But some numbers are associated with a unit. The unit of money can be dollars and cents; the unit of weight can be pounds and ounces; the unit of length can be feet and inches; and so on. The units have to be handled by the programmer. A computer cannot handle units unless it is programmed to do so.

Computers subject numeric data to arithmetic manipulation, and it is possible to perform all arithmetic operations on numeric data. Numeric data is further classified into the following types:

1. *Integers*: Integers are whole numbers without any fractional part. Except in counting, the real world uses fractional numbers. Integers are used in representing age, income, addressing memory locations both in RAM as well as on secondary storage, and so on. In reality, the age of a person remains an integer just for a day, on the birthday. After that, the age would have fractional part. When integers are subjected to the arithmetic operation of division, it may result in a number with a fractional part. Some programming languages allow for two types of integers, namely, the short integer and the long integer. Two bytes are allocated for short integers and four bytes are allocated for long integers. The first bit of the integer is reserved for sign, positive or negative, and the remaining bits are used to store the number. A short integer would have a

maximum value of 32,767 (2^{15}–1). A long integer would have a maximum value of 2,147,483,647 (2^{31}–1). When we use integers, we ought to be sure that the number would never result in fractional part.

2. *Real numbers*: Real numbers are those that have a fractional part. Examples are 10.5, 21.25, 100.35, and so on. The real numbers are stored inside the computer in a special manner. They are stored as the mantissa and the exponent along with a sign. For example, the number 100.25 is stored as (+0.10025, 3). That is the number is 0.10025 multiplied by 10^3 resulting in 100.25. The real numbers are also referred to as floating point numbers. Real numbers are usually stored as single precision and double precision numbers. The precision denotes the number of significant digits the number can store. We should be careful in using double precision numbers, as they take up significant amount of RAM and storage space which is twice that of a single precision number.

   a. Single precision numbers are those that are allocated a minimum of 4 bytes. Presently, most computers allocate 8 bytes to single precision numbers. Single precision number with 8-byte allocation can handle up to 20 significant digits. When the language does not explicitly specify the precision, the default precision is single precision number.

   b. Double precision numbers are those which are allocated a minimum of 8 bytes. Most modern computers allocate 16 bytes to double precision numbers, especially those computers that are used in scientific and mathematical applications. Double precision numbers with 16-byte allocations can handle up to 39 significant digits in the number.

3. *Dates*: Dates are a special category of numeric data. They contains three distinct parts, namely the day of the month, the month, and the year with rules for dependency of the date on the month with respect to the maximum value a date can be. Dates are two types, namely, the short date and the long date. The short date would consist of only the date, month, and the year. The long date would consist of time in hours, minutes, and seconds in addition to the date, month, and the year. The actual storage of a date depends on the computer, but most store the date as the number of seconds from a reference date such as 1900-01-01 00 hours 00 minutes 00 seconds. When displaying the date, they convert it to the date format that human beings are used to see.

4. *Time*: Time is also a special type of numeric data. It contains three numbers for the hour, the minute, and the second. The minute and the second would have a maximum value of 59, then roll out to 00. Hour would have a maximum value of 24. Often the time needs to be displayed with a suffix/prefix of AM or PM and also in a 12-hour format or 24-hour format.

5. *Currency*: Currency is also one special number whose fractional part is restricted to 2 digits after the decimal point. Financial applications use this data type, as the digits after the decimal point are always two.

6. *Counters*: This data type is integers and is used to count the number of iterations a set of program statements is executed inside a program.

7. *Auto incremented numbers*: These are used in database tables to form a primary key for the table when a data item cannot be unique in the table. We will explain the primary key in the next chapter on data storage. The feature of this data type is that it will be automatically incremented whenever a new record is inserted in the table.

We need to note that numeric data is the main data, as the main purpose of computers is to process numeric data. Most common numeric types of data used in computers are detailed earlier. Some computers can have additional types to make life easier for the programmers.

Special Data Types

There are a couple of data types that are used for special purposes in programs, especially in developing system software. These are:

1. *Arrays*: Arrays are tables of data. Usually, arrays are used in solving mathematical problems in matrix algebra. Arrays can be single dimensional, that is, there is only one row of data, or two-dimensional, containing both rows and columns. Most programming languages restrict the use of arrays to numeric data, but some programming languages do allow use of character-type data in arrays. In computers, an array is a contiguous chunk of RAM. The amount of RAM allocated to an array depends on the data type used in the array. If we declare a single dimensional array of six integers, then the computer would allocate a contiguous chunk of RAM that can accommodate six integers. If we declare a two-dimensional array, say, a 3 × 4 array of integers, then the computer would allocate a contiguous chunk of an array to accommodate 12 integers. Arrays are used in both commercial software as well as system software development. Arrays have to be declared separately like other data types, but they need the following additional information besides the name of the array:

    a. Type of data used by the array

    b. Dimension of the array; that is, the number of rows and columns contained in the array

2. *Pointers*: Pointers are a special type of integer data. Pointers can hold the address of any location inside the RAM. So, the size of a pointer type variable is fixed so it can hold the maximum address of any location in the RAM. Usually, the size of a pointer would be 4 bytes. Using pointers, we can access any location in the RAM and manipulate its contents. Application programs are not allowed to access any RAM location, as it would violate the security of the execution environment. In application programming, pointers are used for handling arrays, especially in the C family of programming languages. In system software programming, we need to handle allocation, deallocation, and manipulation of all the RAM, and pointers are used to manipulate RAM as desired by the programmer in the development of system software.

3. *Union*: Union is a special data type used in the C family of programming languages. It allows the same amount of RAM to be addressed using different variable names as well as different data types. It is perhaps the only data type that allows storing both numeric as well as character data and allows the data to be manipulated as both numeric and character data. Of course, it is incumbent on the programmers to store numeric data in a union before handling it as a numeric variable and to store character data in it before manipulating it as a character

variable. This is very useful in real-time software development as there is a premium on the amount of RAM available.

4. *Boolean*: This data has only two states, namely, 0 (true or yes) or 1 (false or no). It is akin to a flag that either waved or held steady. This type of data is sometimes referred to as flags. Most computers dedicate a bit for this data, but as memory comes in chunks of bytes, one byte gets used for this data type. This data is used to set the state when a set of circumstances occurred. For example, as you type a document, a flag is used to denote that the document is changed. When you press the save button, this flag would be checked and if it is set to true (or false, depending on the programmer), the document would be saved. If you press the save button immediately after a save and without changing the text in the document, this flag will remain false (or true, depending on the programmer) and the save action would not be initiated. Flags are also used to denote the end of file conditions, existence of further records in a database table, and so on. Boolean data types are used in all types of software, including business software and system software.

Data Classes

Basically, there are two data classes, namely the local (dynamic) and global (static). The distinction surfaces when the program contains subprograms.

Local Data

Data declared in a subprogram is usually local data. That is the value stored in a local variable that is used only by the subprograms in which it is declared. It is not available to be manipulated by the main program calling the subprogram. The variable, declared as local to a subprogram, is released to the operating system for allocation to other programs when the execution of that subprogram ends. If we want the value stored in the variable to be available to the main program or other subprograms, we must declare it as a global variable. A variable declared in a subprogram is by default a local variable unless it is declared as a global variable.

Global Data

A variable declared in the main program is available to all the subprograms called by the main program for manipulation. It need not be declared as global, but if we declare a variable in a subprogram and want it to be available for manipulation by other subprograms or the main program, we need to declare it as global. A variable contained in a higher-level program is available for manipulation by a lower-level program, but the reverse is not true. A variable declared as global would not be released until all the programs in the main program, including the main program, are closed. A global variable ties up the RAM until all subprograms and the main program are closed. Therefore, it is better to declare variables as local to reduce the burden on RAM. If more RAM was tied up by global variables and other programs need RAM, virtual RAM has to be utilized with all its attendant disadvantages.

Use of Data in the Programs

The primary purpose of computers and computer programs is to process data. To use data inside the programs, we need to declare data that is proposed to be used in the program. To tell the computer we will be using data of a certain type in the program, we declare a variable; thus:

1. A variable is a name for our reference representing the data we propose to use inside the program.
2. The rules for naming a variable vary from one programming language to another programming language. We will discuss more about naming variable in the coming chapters.
3. A variable is associated with a specific set of data to be used in the program.
4. Depending on the data type, the computer allocates the amount of memory (RAM) when the program is loaded for execution into the computer.
5. Whenever we use that variable in the program, it refers to the allocated RAM.
6. The allocated amount of RAM remains allocated as long as the program is under execution in the computer.
7. When the program completes execution and is removed from the RAM, the space allocated for the data will also be released for allocation to other programs.

Thus, to use data in the programs, we declare variables for all the data items we propose to use in the program. Then we keep reading data into the specified locations in the RAM and process it as long as the data exists.

Next, we will see how data is stored in the computers for storage and usage.

3

Data Storage and Retrieval

Storage of Data

To process data, we need to supply it to the computer. We can do so item by item when asked by the computer. We get a prompt on the screen and we input the data using our keyboard. This procedure is alright when we process small amounts of data and in classroom hands-on sessions, but not in the real world of business and government. We need to process bulk amounts of data in the real world. It is just not possible to input bulk data using only keyboards. Therefore, we need to store data.

We will be storing the data inside the RAM when the programs for processing the data are being executed. During program execution, only one set of data is stored inside the RAM. Bulk data is stored on secondary memory or storage. Secondary storage is magnetic media, optical media, and solid-state memory.

In the beginning, we used to store the data on punched cards. Then we moved on to magnetic tapes. Then we moved on to magnetic disks. Now we have optical disks that are now referred to as CDs (Compact Disks) and DVDs (Digital Video Disks). These optical disks use laser (Light Amplification by Stimulated Emission of Radiation) technology. While optical disks have made giant strides in technology, they are yet to replace the magnetic disks as the primary media for data storage. First, the cycles of read and write, on optical disks have not yet achieved the reliability of the magnetic disks. Second, the capacity of the optical disks is not large enough to replace the magnetic disks. Third, the number of times the read and write can be performed on the optical disks is not high enough to match the magnetic disks.

Solid-state memory (flash drives, pen drives, USB disks) is looking very promising, and they can very well replace the magnetic disks as the primary medium for storing data for the long term. They are now capable of holding data up to 250 GB, which makes it feasible to be used in personal computers to begin with. Solid-state memory is the primary reason behind having handheld computers, which are referred to as tablet computers.

Magnetic tapes were the primary choice for storing data until the 1970s but were superseded by magnetic disks. Since being replaced by magnetic disks, tapes were used as backup media to keep backups of data and programs. Even today, large-capacity magnetic tapes are used to store 70 GB or more on data cartridges that are basically quarter-inch wide magnetic tapes kept securely inside a cartridge.

Magnetic Disks

Magnetic disks have been the primary choice for the storage of data for some time now. Magnetic disks are now hermetically sealed and mounted inside the computer to store data and supply data and programs to the computer. Magnetic disks are capable of holding terabytes (one terabyte or TB is equal to 1000 GB) of data. The time to access any location on the disk is also the lowest next only to the RAM. Solid-state drives are competing with magnetic disks in terms of access time but not in the capacity to hold data. Magnetic disks are now referred to as hard disks or Winchester disks. They are called hard disks as they were used at the same time floppy disks were used. While floppy disks were made with polyester film, these disks were made on metal platters. So, polyester disks were referred to as floppy disks and these, being on metallic platters, were referred to as hard disks. Though floppy disks are passé, the name "hard disk" continues.

The magnetic disk is cut into many tracks for holding data. Each track is then divided into number of sectors. The first track is numbered as 0 (zero) and the numbers continue upwards. The sector is the smallest element of the disk that is addressable. In some computers, the sector would be numbered from 0 (zero) upwards within the track. That is, there will as many sectors numbered zero as there are tracks. This necessitates supplying two addresses, namely the track number and the sector number, to locate data on the disk. In some computers, the sectors are numbered from zero upwards without reference to the tracks. In this method, we need to supply only one address, the sector number, to locate data on the disk. Presently, most computers use the latter practice of numbering the sectors; the first sector of the disk is numbered 0 (zero) and the last sector would have whatever number it comes to depending on the capacity of the disk. Each sector holds one block of data. Most computers use 1 KB (1024 bytes) of data as one block or one sector.

The block is the minimum amount of data allocable to files whether we use it entirely or not. When we have more data than can be allocated in one block, the disk space is allocated in multiples of blocks.

When we format a disk to make it usable in the computer, the computer writes a table on the disk. This is referred to as VTOC (Volume Table of Contents), or in PCs, it is called the FAT (File Allocation Table) or a similar name. This table usually contains:

1. The file identification number that usually starts at zero and is numbered upwards.
2. The name of the file.
3. The identification number of the block where the first block of data of the file is stored.
4. It may contain additional information about the size of the file, the addresses of the other blocks of data of the file, the address of the last block of information, and so on.

The disk would also have one more tables in which the block address and to which file it is allocated are written. Initially, the allocation information will be blank. As the disk gets filled up, the allocation information would be filled in at every allocation.

Records

Before we move on to data files, we need to understand records. A related set of data items is one record. For example, let us assume a pay roll situation. An employee has an ID, a name, his basic rate, number of payable hours, other allowances, and deductions. In this manner, all the employees in the organization have similar information. Now, the data would be organized for each employee, for all the employees. If we assume a table, each row will contain the information of one employee and all the rows put together would hold the information of all employees. Each row can be viewed as a record. To summarize:

1. A data file contains several records.
2. Each record contains several data items.
3. Each data item would contain one specific attribute of an entity. Each data item is referred to as field.

Data Files

Data is organized in files in the computers. Why were they called files? Initially, decks of cards were used to store data. When you punch data onto the punch cards, and the cards are bunched into a deck, it would resemble a paper file with papers. Then the data moved on to magnetic tapes, with each tape holding one class of data or, in other words, one file. Each tape had one file of data. Though the data moved onto magnetic disks, the name "file" stuck, as people are familiar and more comfortable with that name. Even though IBM began using the name "data set," which I think is more appropriate to describe a set of data, "file" is definitely more popular in the computer and data processing world.

Although the advent of Database Management System (DBMS) has moved the mainstream data storage from plain data files into the tables of the databases, data files are still used in a significant manner. Therefore, we need to learn about data files before we move on to databases.

Traditionally, files were organized as sequential-access data files and with the introduction of magnetic disks, random access data files came into existence. Then, to combine the advantages of both file types, indexed sequential access data files were introduced. We will discuss these files in greater detail now.

Sequential-Access Data Files

Initially all data files were of sequential access; that is, data had to be read record by record and it was not possible to access any required data directly without accessing all the preceding data records. When the data was on punched cards and magnetic tape, this was the only way. Before we could get to a desired card, we had to read all its preceding cards. On tape too, the tape runs only in one direction and to get to the desired data, we had to pass on the preceding data. With disks, it became possible to access any desired data directly as the disk was spinning and it is possible to access any track and sector within almost the same amount of time. In sequential access data files, there were two kinds of organization.

1. Variable record length sequential-access data files: In these files, the length of each record can vary. It can happen when the fields of the record have variable length. For example, the names of persons would not have the same number of characters. For example, the name John has four characters, while Harold has six characters. Similarly, the pay rate for an employee can be less than 100, having two digits, whereas another employee can have pay rate more than 1000, having more than three digits. When the fields have different lengths, to distinguish one field from other in the same record, a delimiter is inserted after each data item in the record. Usually, a comma, a space, or a semicolon is used to delimit one field from the other in the record. The advantage is that the record length for each record, and thereby the size of the file, is maintained at its minimum. While this was not a great advantage during the punched card days, it was a great advantage during the initial days of magnetic disks, as the disks did not have the large capacity that is available in the present day. The flip side is that an additional character in the form of a delimiter has to be inserted between adjacent fields in the record, increasing the overall record and file sizes.

2. Fixed record length sequential-access data files: In these files, each field has the same length. If a specific data item has less than the allocated length, the extra length is wasted. In numeric fields, leading zeroes are inserted to pad the field to its full length, and in character fields, blank spaces are suffixed to the field as needed to pad the field to its full length. With fixed length fields, the need for delimiters is eliminated. This was more or less the practice in the days of punched cards and magnetic tapes. With the advent of the disks, variable record length sequential access files have become the norm. Still, fixed record length is used in mainframe computers even in current times.

Now there is one other variant in the sequential access data files. It is referred to as the line sequential files. This is mostly used in PC (Personal Computer) applications. In fixed record length files, no delimiter is used between adjacent records. The record sequence is determined by counting the number of characters and dividing the count by the record length. In variable record length files, the record sequence is determined by counting the number of field delimiters and dividing the count by the number of fields in the record. It works perfectly with files on magnetic tapes, but the files on the disk have to use a mechanism to distinguish records from one another. They use the carriage return (ASCII character #13) character and/or line feed (ASCII character #10) character. Windows-based PCs use both the characters and UNIX-based computers use only one of them. These files are referred to as Line Sequential files.

Sequential-access data files offer the best economy in terms of disk space and are still being used. In disk-based sequential access data files, the disk space is allocated contiguously as far as possible and the file is continued in the free blocks available on disk. In this mechanism, a record may span two or more blocks depending on the record length. Sequential-access data files are still used in a big way, especially in mainframe computers and where bulk data needs to be processed.

Random-Access Data Files

The advent of magnetic disks made it possible to access any record in the file with almost the identical access time. This has led to data being organized as random-access data files. In random-access data files, any record in the data file can be accessed by specifying the record number relative to the first record in the file. To facilitate random access of records in the file, a record always began on a new block on the disk. If a record length was less than one block, the remaining space on the block was wasted. If a record spanned more than one block, the remaining space in the second/last block was wasted. Thus, in random access data files, fragmentation of space caused wastage of disk space. Purely random-access files did not last long and were quickly replaced by indexed files. Random files were very efficient in locating a specific record based on a key value such as a name or an identification number.

Indexed Sequential-Access Data Files

In order to combine the advantages of random access files in their ability to quickly access any desired record and the efficient utilization of disk space by the sequential access files, the industry came out with indexed sequential-access data files. These were popularly referred to as ISAM (Indexed Sequential-Access Method) files. In this method, an index to the records was maintained as a separate sequential-access data file. This index file contained just two values for each record, namely, the value of the key data item and its relative location in the file. Whenever a record needed location, the index file was searched first sequentially to locate its relative location in the main data file and then, using the relative location, the desired record was located and retrieved for use. This had greatly reduced the time for locating any desired record. The extra space needed for the index file was much less than the space wasted in the fragmentation loss of the disk-space blocks.

ISAM files are still used, especially in COBOL applications in IBM and other mainframe computers. These are used in bulk data processing applications. IBM names sequential-access data files as ESDS (Entry Sequence Data Set) and index sequential-access data files as KSDS (Key Sequence Data Set).

But maintaining the data files had to be accomplished through programs. The extra effort needed was significant. The data had to be entered offline, that is, on a separate data entry machine by a specialist data entry operator. The entered data needed to be perfect in every way, and 100% accuracy was desirable but never achieved. Correction of data was really laborious. This led to errors in processing. Even if the data entry was 99.99% accurate (that is an error rate was 0.01%), the number of defective records in a million could be as high as 10,000! While the percentage could be low, the absolute number was unacceptably high. The organizations needed significant numbers of employees to handle the complaints of wrong results. Another major disadvantage with data files was that they did not facilitate being used by the employees specializing in their domain. All the transactions were carried out on paper and data was entered offline. The best that computers could do was to supply management information for decision making but could not really make significant impact on the cost reduction in the organizations. To reduce the costs in the organization, the transactions needed to be moved from paper to computer. This required better organization of data than these data files, and that shift came in the form of databases or DBMS.

By the way, these data files are referred to as "flat files" in the current time, as the files are flat and without any control information that exists in the databases. In real-time applications, flat files are the ones used to store data. In machine-control applications, such as the software used in cars, airplanes, rockets, and so on, flat files are used to store data. In firmware, that is, software etched onto a silicon chip, it is the flat files that are used. It is only in commercial applications that databases handle the bulk of the data, but even in them, a few flat files are used. Therefore, serious computer programmers need to learn about and understand flat files and their organization and usage.

Database Management System

The main issue with having data in flat files was that it needed maintenance by programs written specifically for file maintenance. Every time the file structure changed, that is, either a new field was added or a field was eliminated, all the programs needed change, especially in the sections where the file structure was defined. To alleviate this problem, the definition of the file structure was taken out of the programs and stored separately as files. The programs just made references to the definition in those files. But if the file structure changed, the programs had to be recompiled and converted to executable files once again. If the file structure changed to add a new field for the sake of one program, all the other programs that used that file had to be recompiled, even if the changed file structure did not affect them. This not only caused organizations to spend considerable effort in program maintenance, but it also caused errors whenever a program was overlooked during maintenance.

The industry worked on overcoming this problem and came up with DBMS technology. DBMS is a software package with a set of programs that facilitate:

1. Definition of file structure.
2. Maintain file structure including addition, deletion, and modification of fields.
3. Enter data into files (referred to as tables).
4. Maintain data including addition/insertion of new data, deletion of existing data, and modification of existing data.
5. Create and maintain index tables as necessary so that indexed search capabilities are made available for programs in fast data retrieval. Also, update indexes as and when the data changed.
6. Provide a language that is optimized for fast conditional data retrieval that can be called from programs. This has relieved the programmers from the hassle of coming up with the best possible routines to retrieve data efficiently and quickly.

Databases have been adopted by the industry quickly, and in present day commercial applications and those applications that using large volumes of data, DBMS has become an inseparable part of the applications. With DBMS, the programmers need not define the file structure inside the program. They need not even recompile the programs, even if the file structure changed. Another major advantage of DBMS technology is that no programs need development for entering data. DBMS provides data entry facilities. Another major disadvantage of flat files that prevented multiple users from concurrently using the same file was overcome by DBMS. With DBMS, the entire file need not be dedicated to a program. It is adequate to lock one record to the program and other programs can concurrently continue to use the other parts of the table or the database in their applications. This facilitated online data processing by end users of the computers, and the need for specialist data entry operators was eliminated. As users are now entering the data directly, the need for data verification was also eliminated.

DBMS has facilitated storing large volumes of data simultaneously by multiple users, and efficient and quick data retrieval. This has facilitated moving business transaction from registers and papers to computers and paved the way for paperless offices.

Now, DBMS has four levels of organization:

1. *At the bottom-most level is the field or domain*: It is the smallest unit of data and holds data of one specific attribute of the entity. For example, a name is a field and holds the values of names; similarly, pay rate is a field that holds the values of pay rates for different people. A field is like a column in a two-dimensional table.

2. *At the next level is a record or tuple*: It holds the complete information for an entity. For example, in pay roll data, a record contains the pay information for one employee. It would contain information like the employee identification number, name, pay rate, other allowances, deductions, payable number of hours, and so on. A record consists of multiple fields. A record is akin to a row in a two-dimensional table.

3. *A table or relation*: It is related information for an application. It consists of a number of records. In a pay roll application, a master table would consist of master data of employees such as employee id, name, designation, pay rate, allowances applicable to the specific employee, normal deductions, and so on. A pay roll transaction table would consist of employee id, number of payable hours, other earnings during the pay period, deductions during the pay period, and so on. In this manner, there can be other tables in the pay roll application.

4. *A database*: It is a collection of tables that are related together. For example, a pay roll database may consist of employee master table, transaction table, union membership table, income tax table, and other such tables. An organization may have multiple databases such as a marketing database, material management database, finance database, personnel database, and so on.

In the programs, a connection needs to be made to the database and then all the data in the database can be used: any record can be retrieved, any field can be updated, any record can be added or inserted, any record can be deleted. In this manner, any data can be manipulated as required by the organization. DBMS technology has removed the dependency on data for the programs. Now, programs and data are independent of each other and can be maintained separately without worrying about their impact on each other.

Presently, most commercial applications make use of database to handle the data of the applications. Databases evolved from hierarchical databases and network databases to relational database technology. In hierarchical databases, relations could be set from one table to many tables. In network databases, relations could be set from many tables to many tables. In relational databases, relations could be set from one table to just another table. It is a one-to-one relationship.

However, by using interface tables, we can achieve many-to-many relationships in relational technology. Relational databases utilize disk space very efficiently and hence are preferred over the other two technologies. Oracle, SQL Server, Ingress, Informix, MySQL, and Progress are examples of commercially available relational DBMS software. A standard scripting language of SQL (Structured Query Language) was developed for data definition, retrieval, and maintenance. Programmers can call the SQL routines in their programs directly and make use of them for data maintenance and retrieval.

The design of databases and the definition of tables can be performed independently of the software development. Let us now understand a few basic requirements of DBMS packages so that we can effectively use them in software development.

Each table needs to have a primary key. A primary key is a field in a table in which data values are not duplicated. Each data value of the primary key is unique. Without a primary key, a table cannot be defined in a truly relational database. In some relational database systems, a combination key with the values of two or more fields can be combined to form the primary key, but most systems require one field to be defined as a primary key. In certain cases, it may not be feasible to set aside any one field as a primary key. In such cases, we usually define an additional field and fill it with auto incrementing numbers in order to have a primary key.

As some perceive, DBMS does not completely eliminate the duplication and redundancy of data across multiple tables. What they really do is control the redundancy. To set a relationship between two tables, a field must be common between those tables. It is referred to as the secondary key. While the primary key cannot have duplicate values in the table, secondary keys can have duplicate values. Incidentally, the secondary key in a table ought to be the primary key in another table with which the current table is related. A relation is set between the primary key of a table and the secondary key of a different table. A table can have only one primary key, but it can have multiple secondary keys.

The design of a database is a large enough subject to merit a book in itself. Here I am including a brief explanation for your understanding of the basics. Usually, in organizations specializing in software development, it is common to dedicate a database specialist to design the database and develop routines for efficient data manipulation that are used by the programmers in their programs. Here are the steps in the design of the database:

1. First, collate all data items used in the application.
2. Next, logically divide the data items into logical groups.
3. Then, carry out a process referred to as normalization. Normalization is a systematic process of elimination of repeating data items into logical groups. It is usually carried out in three stages, and the third normal form would have a primary key and all other data items in the group would not have duplicates.
4. Each of the logical groups in the third normal form would then be defined as a separate table.
5. Most DBMS packages would have inbuilt facilities to build indexes and set relations automatically.

Each package would have different rules for naming the tables and fields, as well as data types for defining data, but most DBMS packages would support the data types described earlier in this chapter. All the DBMS packages implement the standard SQL language with some extensions and a few adaptations of their own.

Depending on the type of application you are developing, the type of data storage and retrieval needs to be determined and used in the development. Chapter 2 introduces you to the two basic types of data storage and retrieval, namely, the flat files and the DBMS packages. Now we are ready to understand what computer programs are.

4

Introduction to Computer Programs

Introduction

What is a program? Merriam Webster's dictionary defines "program" for contexts other than computers as "a plan of things that are done in order to achieve a specific result," and as "a thin book or a piece of paper that gives information about a concert, play, sports, games, etc." If you look at a program sheet, you would notice that it lists all the items that needed performance, and each of the items in the list are arranged in the chronological order of their performance. Usually, the program is adhered to in the performance of activities to achieve the desired result.

Merriam Webster's dictionary also gives another definition that is pertinent to computer programs: "a set of instructions that give information that tell a computer what to do."

Standard 610, Standard Glossary of Software Engineering Terminology of IEEE (Institute of Electrical and Electronics Engineers) gives this definition of computer program: "a combination of computer instructions and data definitions that enable computer hardware to perform computational or control functions." It is also pertinent to note IEEE's definition of software: "computers programs, procedures, and possibly associated documentation and data pertaining to the operation of a computer system."

Wikipedia defines a computer program as "a sequence of instructions written to perform a specified task with a computer."

Oxford dictionary defines program as "a series of coded software instructions, to control the operation of a computer or other machine."

I think these definitions are adequate for us to begin understanding what a program is in the context of computers and software. One interesting aspect to note is in the definition from the Oxford dictionary. It included the phrase, "or other machine." Now we have software controlling rockets, airplanes, cars, washing machines, and even children's toys! But, these machines have a processor in them to execute the software and then pass on appropriate signals to the machine. The processor may not be as powerful as the ones in computers, but they are processors. So, in order to determine realistically if these processors inside machines are computers or not, we need to look at a universally and undisputedly accepted definition of "computer." Unfortunately, such a definition eludes the world. That is why, perhaps, Oxford dictionary included "machine" in its definition, as it perceived the processor inside the machines not as computers. Others perceived that the processors inside machines are computers and did not include "machine" in their definitions.

I prefer the definition of IEEE and I wish to extend it thusly: "a combination of computer instructions and data definitions *arranged in their chronological sequence of execution* that enable computer hardware to perform *input/output,* computational or control functions."

The italicized parts in the earlier definition are my insertions. Now, adopting this definition of a computer program, we can move forward.

Computer programs are in three states, namely, the source code, the object code, and the executable code. Source code is the program written by programmers in a computer programming language. It is humanly readable and understandable. Object code is the compiled source code program. It is in a machine readable and understandable form. Executable code is the object program linked with code libraries and is ready to be executed by the computer. Basically, the object code and executable code are similar. Object code is the compiled program that is then transformed into executable code by linking it with the applicable libraries.

Many synonyms are used to refer to computer programs. The most common synonyms are macro, script, agent, trigger, procedure, function, applet, servlet, app, bot, and routine. Perhaps there could be more, and more would be on the way. While each of these terms has a definition, they are all basically programs. Programming covers all these types of programs.

Components of a Program

A program would have the following components, normally:

1. *Headers*: These are statements placed above the program code and can be used in the entire program.
2. *Program beginning and ending*: Every program has an identifiable beginning and an identifiable ending. These contain specific sets of actions to be performed when the program begins execution and ends the program.
3. *Data definitions*: These are definitions of the data proposed to be used during the program execution. Each definition results in the allocation of RAM as required.
4. *Input operations*: These operations bring in data from the outside world into the RAM for processing.
5. *Output operations*: These operations deliver the processed information to the outside world.
6. *Computational operations*: These operations perform mathematical operations and produce results.
7. *Decision making operations*: These operations make programmed decisions and control the program flow.
8. *Program documentation*: These statements explain the logic of the program statements, which assist other programmers in maintaining the program.
9. *System calls*: These operations call the services provided by the operating system and utilize them in the program.

We will discuss all of these in subsequent chapters.

Program Statements

The program consists of program statements. In the earlier days, the card consisted of one statement and, as a punched card could accommodate 80 characters, a statement was restricted to one card. Continuation of the statement in another card was permitted in special conditions but was used sparingly. Later, when the VDU (Visual Display Unit) was introduced, the programming shifted to VDUs. They also were designed to support 80 characters per line, so the same rules continued for the line and statement. But modern computer screens can display more than 80 characters per line and can even scroll to the right side to accommodate more characters. That is why modern programming languages support up to 255 characters per line, but the good practice of writing programs suggests including as many characters as the display unit can accommodate in a line without the need for scrolling the screen horizontally.

A line ends with an LF (Line Feed—ASCII character #10) character and/or CR (Carriage Return—ASCII character #13) character. It is easy to count the number of lines in a program as we just need to count the LF/CR characters.

In today's programming environment, a statement can span across multiple lines. Some languages need some indication that the line is a continued statement, but many development environments do not need this. Some languages use a statement terminator character such as a semicolon (;). We can write any number of lines as required between two successive semicolons and all those lines would be treated as a single statement by the computer.

Source Code, Object Code, and Executable Code

Computer programs are referred to as code. Why the term "code?" Computers use a language different from the one used by human beings. It is composed only of zeroes and ones. Using those two digits, all instructions are codified. When we write programs in a language understood by human beings, we need to follow certain rules so that they can be converted into the language understood by the computer. So, when we write programs, we cannot write using free-flowing language but need to adhere to certain rules and, at best, we can write approximate to human language. That is why we call the computer programs computer code.

The computer programs written in humanly understandable language are called source code. It means the program is the source of all the programs that follow. Source code is the program written by a computer programmer. Then this program is translated into a program that is understood by the computer.

The source code (or source program) is translated into computer language. While the computer language is zeroes and ones, it actually consists of instructions that run the computer. These instructions are actually numbered. Each of those numbers represents one specific action to be performed by the computer. These are called as opcodes (Operation Code). How many opcodes would be there in a computer? It depends on the designers who designed the computer. Obviously, a microcomputer would have less number of opcodes than a mainframe computer. Depending on the number opcodes (or instructions) a computer supports, the opcode would use the bits. A four-bit opcode can support 16 instructions and a 8-bit opcode can support 256 opcodes. From now on, I will use "instruction" where the term "opcode" needs to be used so that we have easier readability.

When the source code is translated, it is translated into object program, or object code. It simply consists of computer instructions and the data definitions. So, object code is the source code translated into computer instructions.

Now, the data definitions need to be allocated RAM. As we noted earlier, the data is of different types. So, each of the defined data items would be allocated RAM in appropriate amounts. An integer would be allocated two bytes, a single precision real number would be allocated four bytes, and a character string would be allocated the number of bytes equal to the number of characters it needs to hold. In this manner, RAM would be allocated to all data items defined in the programs. But, at this point in time, it would not be known where the RAM would be allocated. So, the computer would allocate the total amount of RAM needed, the number of items needed, and the address of each item from the first item allocated. Actual locations would be decided when the program is submitted to the computer for execution. This is referred to as the "relative addressing." Each address is relative to the first address.

Let us assume that the program needed 100 bytes for four locations. The first item begins at location 1 and extends to 4, the second item begins at 5 and extends to 60; the third item begins at 61 and extends till 92; and the last item begins at 93 and ends at 100. Now, suppose the allocation of RAM begins at location 102,401. Then the address of the first data item is 102,401; the address of the second data item begins at 102,405; the address of the third data item begins at 102,461; and the address of the fourth data item begins at 102,493. This is just an example to give you an idea about memory allocation and relative addressing. Now, all data items are allocated.

Every programming language makes use of the services of the operating system of the computer. These services consist of receiving inputs and delivering outputs, using RAM, and using the CPU. Apart from those services of the operating system, certain other features like the mathematical operations and commonly used functions are provided by the programming language. Now the object program is equipped with memory locations and linked to the services and common routines, making the program ready for execution by the computer. Now, this code is referred to as the executable code.

Summarizing our discussion:

1. Source code is the computer program written by the programmers in a language understood by human beings.

2. Object code is the computer program that is translated from the source code into a language that is understood by the computer.

3. Executable code is the program that has been allocated with RAM for all the data items defined in the program and is linked with all the services and programs made available by the operating system and the program libraries.

A program library is a set of programs made available by the manufacturer of the computer, the developers of the operating system, and the developer of the programming language used for writing the computer program.

Here we need to understand the term "compiler." A compiler is a computer program in itself that performs the functions as follows:

1. It checks the program for adherence to syntax rules specified by the specific programming language. When errors are detected, it lists out the errors so that they can be corrected.

2. When the program is completely free of syntax errors, it translates the program into object code.

3. It computes the amount of RAM required for all defined data items and allocates them using relative addressing.

4. The final output of a compiler is the object program that can be used to link to its concerned libraries to prepare the executable code.

Computer Programming

What is computer programming? *Computer programming is a set of activities including writing a program in a programming language, then compiling it and linking it to the relevant libraries to prepare the executable code, and then testing the program to remove all errors including syntax errors, logical errors, and computational errors lurking inside the program.*

Before the advent of the IEDs (Interactive Development Environments), these were the steps in writing and preparing the executable programs and then executing them:

1. *Use a text editor program to enter the program*: This is the first step of writing a computer program. In the earlier days, programmers used to write the program on a graph paper with one character per each square. Then that was punched on the punched cards by specialist data entry operators using card punch machines. But, with powerful editors becoming available, the programmers themselves were typing out the program into the computers using the text editors. Editors allow entering the required text, making corrections, deletions, copying, and pasting, and all such text-editing facilities. The programmers just need to adhere to the syntax rules prescribed by the specific programming language in which the program is being written. The programmer also needs to write the program in such a way that the desired results are obtained when the program is executed with relevant data on the intended computer.

2. *Compile and debug for syntax errors*: Once the program was written, the programmer submits the program to the compiler. The compiler checks the syntax and enumerates the errors onto a printer or a screen. Then all the syntax errors enumerated by the compiler need to be rectified by the programmer using the text editor. The process of removing errors from a program is called "debugging." Once all the errors are removed, the program needed to be resubmitted to the compiler once again. The program may go through multiple iterations of compiling and debugging until all syntax errors are eliminated from the program and the object code is generated by the compiler without any errors.

3. *Link the object code to the required code libraries*: This step is achieved using a linking program that scans the program and extracts the names of the code libraries the program invoked and then links the program to the invoked libraries. In the earlier days, the code of the code libraries invoked in the program was appended to the program's object code. In the present days, the code libraries are DLLs (Dynamically Linked Libraries). They need not be appended to the program code. They will be loaded dynamically at the time they are called during the execution of the program. The linking program ensures that all the invoked libraries are indeed present in the specified directories. With the successful completion of this step, the executable code is ready.

4. *Execute to ensure that the program is working*: Now we need to execute the program. This involves invoking the program. In the earlier days, it involved typing the name of the program at the command line along with any arguments that are necessary to run the program. In the current times, we select the program from a drop-down menu or double click an icon on the desktop of the computer screen. When we invoke the program, the program is loaded into the RAM and its execution begins. It will display a screen, a message, print, or whatever the program is expected to do. Sometimes, the program may contain such logical errors that the program itself cannot run. This step ensures that the program is indeed running and can be tested to ensure that it is doing what is expected of it.

5. *Test, test data, accuracy of computation, retrieval of right data*: Every programmer is expected to test the program written by him/her thoroughly to ensure that the program is doing what it is expected to do and is not doing what it is not expected to do. This is referred to as "self-testing." Self-testing may not be as thorough as the testing performed by a specialist tester. Still, the programmer is expected to test thoroughly so that no errors can be uncovered by the specialist tester. After all, the programmer is the one who did all the programming and is the best person to know what it can or cannot do. Testing forms part of software quality assurance and is important enough to merit a separate chapter. We will discuss more about it in the subsequent chapters.

Now, all these need not be separate steps. With the advent of IDEs, all these can be performed from within the IDE. An IDE allows for entering the program code, editing it, compiling it, running it, testing, and debugging it as well. Each supplier of the programming language supplies an IDE tailored to their programming language. Commercial off-the-shelf IDEs are also available for use in the software development.

Debugging has become a very popular term in software development circles. The term came about as follows. The first practical electronic digital computer was ENIAC (Electronic Numerical Integrator and Calculator) was built using many valves and relays. One day, the computer stopped working and the technicians worked for a long time to find the problem and rectify it. On the night shift, a technician found a moth stuck between the contacts of a relay. She removed the dead moth and cleaned the contacts and then the computer began working again. In the shift log she wrote, "The computer was 'debugged' and it is now working," or something similar to that. From then on, the term "debug" became very popular and entered the dictionaries. Debug has come to be understood as removing defects from anything.

How Does a Computer Execute a Program?

When we know how a computer executes a program, we would be able to write programs that would be executed efficiently. The present-day computers are referred to as "stored program" computers or as "Von Neumann architecture" computers. That is, the program is stored in the RAM while it is under execution. Von Neumann was the computer scientist who defined this concept and the architecture of the present-day computers is known as Von Neumann architecture.

When we invoke a program for execution on a computer, the following activities are performed:

1. The computer assigns an identification number to the program. This number is used by the computer to refer to the program while it is being executed. A program under execution is referred to as a process. The id assigned to it by the OS is referred to as the "process id." Some computers allow us to change the execution priority of a program using this process id, or PID.

2. The operating system of the computer maintains a table inside the RAM, generally referred to as "process table." It would contain, for all processes, the process id, its state, the beginning address of the RAM where the program is located, the address of the next program instruction to be executed, and the beginning address of the data items in the RAM.

3. The computer assigns the process id of the new program at the end of the table. The state of the process is set to "wait," waiting for allocation of RAM. The other states would be "ready" and "running." A process in the ready state would be in the queue for assigning it to the processor for processing. A process is kept in the wait state if some input/output operation is needed by the process. The process is in running state when it is being executed by the processor.

4. Then the operating system allocates RAM for the program code and the data items. The details of the requirements are taken from the executable code. The beginning addresses of the program code and data are entered into the process table.

5. Then an operating system program, referred to as "loader," loads the program code at the allocated location and sets the process state to "ready."

6. The operating system processor management module allocates a time slice to the process in a round-robin manner.

7. Then CPU is allocated to the process. The CPU then executes the program instructions one after the other. The CPU removes the process under three conditions:

    a. When the time slice allocated to the process expires. Usually the time slice allocated to a process is 100 milliseconds but could vary from operating system to operating system.

    b. When the CPU encounters an I/O (Input or Output) operation. When such an instruction is encountered, the CPU instructs the device manager module to obtain the data from the specified input device or deliver the output to the specified device. The devices input or output the data as instructed by the CPU. At each iteration, we need to note that only one set of data items is either input into the allocated RAM or output from the RAM.

    c. When the CPU has to allocate time to a process that has a higher priority than the process under execution. This situation is called an "interrupt." When an interrupt is placed on the CPU, the CPU deallocates the process under execution and attends to the interrupt.

8. The process is then set to the wait state and sets the address of the next instruction to be executed in the process table.

9. The process continues to be in the wait state until the I/O operation is completed. When the I/O operation is completed, the state of the process is set to "ready."

10. If the process is deallocated only because its time slice expired, then the state of the process would be set to "ready" when it is deallocated.

11. All the processes in the "ready" state would be in the queue for the allocation of CPU. The CPU would be allocating time to each of the processes in ready state and executes process by process in a round-robin manner.

12. When all the instructions in the program are executed and all the data is processed, the program would be removed from the RAM, all data files and databases opened for the program are closed, and the process is removed from the process table.

This is the manner in which the computer executes the programs.

Programming Styles

We have noted that a computer program is a sequence of statements arranged in their chronological sequence. The computer executes the first instruction in the program and then it comes to the second, then the third, and so on. The computer execution goes like a waterfall from the top of the program to the bottom of the program, executing one instruction after the other. Of course, by inserting control statements, we can make the computer skip the order of execution and execute a set of other instructions. These statements are referred to as branching statements (meaning that these instructions cause diverting the execution to another branch of the program) and control statements (as these instructions control the flow of execution).

When we write programs in such a way as to flow freely from top of the program to the bottom of the program, the style is referred to as waterfall programming. We have to note that no program can be written completely avoiding some sort of decision making and branching based on the outcome of the decisions. The experience had shown that the branching statements in this style of programming branches off the execution to undesirable locations and produces unexpected, inaccurate, and undesirable output. This has been especially true in the case of very long programs exceeding a thousand lines of code. So, this sort of programming is more or less shunned. We use this style in a limited manner, especially in very short programs like macros, scripts, functions, and subprograms.

This style is superseded by what is popularly referred to as "structured programming." The following guidelines define structured programming:

1. There is one main program.
   a. It is short in length.
   b. It contains the beginning and the ending of the programs.
   c. It does not usually perform any function except calling subprograms.
   d. It contains calls to subprograms. It is usually a waterfall program, but can contain decision-making statements that are used to select a subprogram.
2. Each subprogram would perform, usually, one single function.
3. Each subprogram would be short in length. While there is no fixed length specified for such subprograms, usually each would be limited to 50 lines in length. Longer subprograms are used when the logic of the function to be achieved requires longer programs.

Now with GUI (Graphical User Interface) becoming the norm for all software, structured programming has been more or less implemented across the applications. In the GUI, each of the controls has a number of events associated with it. They include gotfocus, lostfocus, click, doubleclick, and so on. Each event needs to be programmed separately. Thus, each event is controlled by a subprogram. Structured programming is inbuilt in the GUI applications.

Readability of Programs

When we write programs, we need to keep in mind one thing and that is that the program could be running for years and it would need maintenance. The programs written in 1960s and 1970s are running even today. The programmers who wrote them retired long ago, and the programmers who are maintaining them are today's youngsters. Therefore, while we write programs, we need to ensure that the programs are written in such a way that it is easy for other people to read and understand them. The following guidelines help us in writing programs that are easily understandable to others:

1. We write each line such that the reader need not scroll horizontally.

2. When a line is subordinate to the preceding line, then we begin it with an offset. That is, if the principal line is starting at position 1, we begin the subordinate line at an offset of one tab-space. Here is an example:

   Here is the first principal line. It begins at position 1

       Here is the first subordinate line. It is offset by one-tab space

           Here is the second subordinate line

   Now here begins another principal line

3. Most programming languages provide keywords to reduce the program length. Such keywords combine a few lines of code into one line. They allow us to write short programs. But, when something does not work as expected, it becomes very difficult to debug those statements that combine multiple lines of code into a single line. When the program is compiled, both programs (with one line or multiple lines for achieving the same functionality) would translate into the same number of computer instructions in the executable code. Therefore, it is better to be writing long programs instead of short programs using complex keywords. Of course, it takes more time to type in more number of lines, but it reduces the time in debugging and the total time taken is almost the same.

4. When a statement has to be longer than what the screen can horizontally accommodate in one line, it is better to break the statement into shorter statements. Alternatively, we can make use of the statement continuation facility to break it into multiple lines. When we continue a statement on a second line, it is better to offset the next line by a tab space so that it is clear that it is subordinate to the previous line.

We need to note that the hallmark of a good program is that it is maintainable in addition to the main functionality of doing what is expected: produce expected results diligently and accurately.

Introduction to Program Structure

Each programming language usually has a predefined structure for the program. Some languages have a very strict structure. For example, COBOL (Common Business Oriented Language) expects the program in four divisions: the Identification Division, the Environment Division, the Data Division, and the Procedure Division. What can go into each of these divisions is predefined. Similarly, the RPG (Report Program Generator) also has a strict structure with some specific character positions on the line having defined interpretations. Others are not so structured. But in every language, the beginning of the program and the ending of the program have specific statements.

Semantics and Syntax

Every programming language defines certain keywords. Most of these keywords are verbs; that is, they specify some action to be taken by the computer. Consider these examples:

1. *Open*: Open something, such as a database, a table, a port, or a device. This statement tells the computer to begin communication with the specified object. Once the operating system encounters this keyword, it sends a command to the specified object to begin the communication. If the specified object is available and ready to reciprocate communication with the calling program, it responds with an acknowledgement, which means that the object is ready to exchange information with the calling program. Sometimes, the device may send a busy response, which indicates to the calling program that it needs to wait and keep calling for the object until it receives a ready response. Sometimes, when the specified object is absent, no response is received. In such cases, the operating system generates an appropriate error message and sends it to the calling program. Then the calling program needs to interpret the error message and take appropriate action as specified by the programmer. When the proper acknowledgement is received from the object, then the operating system makes an entry in the open objects table and stores all required information about the object so that further exchange of information can take place between the object and the calling program.

2. *Close*: This keyword tells the operating system to close the communication with the specified object. The specified object can be a database, a table, a file, or an I/O device. When this statement is encountered by the operating system, it tells the object that it is now free to interact with other programs, its state is set as "free" in the relevant table, and its entry is deleted from the open objects table of the calling program.

In this manner, there are many keywords for each of the programming languages. How does a programmer make use of these keywords? The programming language defines a set of rules generally referred to as the syntax for the language. The syntax specifies the following:

1. The number and type of data items to be used along with the keyword.

2. Other subordinate keywords that are expected to be used along with the main keyword. For example, in many programming languages, the "If" keyword needs to have the keyword "Then," as well as "Else," following it.

3. The exact arrangement of keywords and data items, including the order in which they need to be placed in the statement.

4. The rules for continuing a statement on the next line.

5. The rules for terminating a statement.

6. The rules to hammock a set of statements into one block.

7. The rules for calling a subprogram or the system services offered by the operating system of the computer.

8. The rules for handling the devices connected to the computer, such as the printer or scanner.

The syntax can have other rules that are specific to the programming language. The rules of the syntax must be adhered to in toto. Any laxity in adhering to the rules of syntax would result in an error during the process of compilation. When a syntax error is encountered by the compiler, the compiler will generate an error message and it would not generate the object code.

To be able to write good quality, and efficient programs in a programming language, one must master the syntax rules of the language besides learning as many keywords as possible.

5

Algorithms and Flowcharts

Introduction

Two questions every beginner to computer programming faces are, "How do I know what the sequence of statements should be? How do I solve the problem at hand by writing a computer program?" The answer to these questions is algorithms and flowcharts. First, let us see what an algorithm is and then understand flowcharting. In order to solve problems our minds use a complex mechanism to arrive at solutions. We really do not have to think systematically. For example, if you ask a fourth grader, "What is the sum of 2 and 4?" he would immediately answer, "6." How did he arrive at it? Perhaps he would tell you that he knows. Did he think about it systematically? No! How exactly our minds arrive at solutions to problems is still an area of research, but computers are not gifted as human beings are with a brain. Therefore, we have to understand automated reasoning. Automated reasoning is building a mechanism to reason in a systematic manner and to arrive at accurate conclusions. Algorithms are one of the most popular tools of automated reasoning.

Algorithm

Merriam Webster's dictionary defines "algorithm" as "a step by step procedure for solving a problem accomplishing some end especially by a computer." Those that have studied mathematics in some depth would readily understand algorithms. Most of the mathematical problems are solved using algorithms.

We need to understand that a computer is a sincere and diligent servant who would carry out all instructions "as specified" but would not correct the defects in our instructions. Your brain has intelligence, but there just is no intelligence in a computer. It cannot apply its "common sense" as it does not have any. Everything needs to be programmed and all instructions need to be provided to it. Let us understand an algorithm with an example.

Let us take two numbers and add them. It is a very simple thing for human beings. If I say, "Add 2 and 4." You would answer, "6," in no time at all. How did you do this? Perhaps you are not really aware how this is done, but surely:

1. You heard the problem.
2. You understood it.
3. You considered the number 2.

4. You considered the number 4.

5. You added them together using the knowledge you already possess.

6. You announced the answer as number 6.

Right? Your brain performs these steps so fast that you are not even aware that those steps were indeed performed by your brain. Now if this is to be performed by a computer, almost similar steps would need to be performed. The steps would slightly change their order though. Here are the steps:

1. Receive the input of first number and store it in RAM.

2. Receive the input of second number and store it at a different location in RAM.

3. Receive the instruction of what to do with those two numbers.

4. Perform the addition using the program for solving mathematical operations existing inside the computer.

5. Store the result at a third location in the RAM.

6. Give the result on an output device.

Let us understand each of the earlier statements in detail.

1. *Receive the input of the first number and store it in the RAM*: Computers have no eyes or ears to automatically know when we supply data. The first number is a data item of numeric type. We store data in our brain and the computer stores the data in its RAM. The supplied data remains in the RAM until the program is closed. The CPU needs all data items to be in the RAM for processing. The CPU does not access input devices directly. Receiving the input from a specified device is carried out by a program that is usually part of the operating system.

2. *Receive the input of the second number and store it at a different location in the RAM*: It needs to receive the second number, just as it received the first number, and store it at a different location. In this manner, the computer needs to receive as many data items as we supply to it and keep storing them at different locations until they are processed and the program is closed.

3. *Receive the instruction of what to do with those two numbers*: Once all the data items to be processed are received and stored in their respective locations, we need to supply the instructions to do the operation with the supplied data items. Notice here the contrast between the human beings and the computer. We give the instructions first and data items next to the human beings, but for computers, it is the reverse. We supply the data items first and then the instructions to process the data items.

4. *Perform the addition using the program for solving mathematical operations existing inside the computer*: The knowledge to solve a mathematical problem is stored inside our brain. For the computers, it is supplied by the program libraries included with the programming language. When an instruction is given to us, we consider if we already have the knowledge to perform it. If we have it, then we do it. Otherwise, we respond by saying, "I do not know how to do it." When the computer comes across an instruction, it searches the included libraries to see if the procedure to solve it exists. If it exists, it performs the operation; otherwise, it generates an error message that it cannot perform the operation and gives it as output.

5. *Store the result at a third location in the RAM*: Once the operation is performed, it needs to store the result inside the RAM. The operation is performed by the CPU, and the CPU is not a storage area. So, the CPU transfers the result to a location inside the RAM. All outputs are taken from the RAM onto the specified devices. The CPU does not interact with the output devices directly.

6. *Give the output on an output device*: Once the result is inside the RAM, the program to deliver the output would begin execution to deliver the result to the specified output device. This program for delivering the outputs to the specified device is usually part of the operating system.

From the earlier explanation we can draw some conclusions for our future use:

1. All data items need to be supplied to the computer through an input device. We had seen in Chapter 1 what the possible input devices are. However, when we are developing the algorithms, we need not be concerned with the specific input device that would be used. We simply say "Read" or "Input." Each input device comes with a software called the "device driver" that is installed and becomes part of the operating system.

2. All data items supplied by the computer to the outside world are delivered through an output device. Each output device also comes with a device driver and becomes part of the operating system on installation. At the time of developing the algorithm, we need not concern ourselves with the specific output device to be used. We can simply say "Write" or "Print."

3. All data, either input data items or output data items, needs to be stored inside the RAM. The CPU can access data items only from the RAM.

We stated that the data items and the results of operations are stored at different locations inside the RAM. Who specifies the location for each of the data items? We do, of course! We do it symbolically though. We use a name for the location and the operating system converts it to a location and uses it. For example, we do not say, "Read 3." We instead say, "Read a number A." That is, we are specifying the address of a memory location symbolically referred to as A into which the input needs to be stored.

Similarly, we specify "C = A + B." We are telling the computer to add the contents of memory location A and the contents of memory location B and store the result in memory location C. Now, we can develop a generic algorithm to add a series of two-numbers, as follows:

1. Read A
2. Read B
3. C = A + B
4. Print C
5. Prompt to enquire if there are more numbers
6. Receive the answer from the user
7. If the answer is "Yes," then go to step 1
8. If the answer is "No," then stop

When this algorithm is fed to the computer and the computer executes it, the following actions are taken by the computer:

1. The first statement causes the computer to receive a number from the specified input device. Of course, in the earlier algorithm we did not specify the input device. It would be specified in the program. In case no input device is specified, it would expect the input from the default input device, which is usually the keyboard. The number received from the input device is then stored in a memory location symbolically addressed as A.

2. The second instruction causes the computer to receive the second number from the input device and stores it at the memory location addressed as B.

3. The third statement causes the computer to take the contents of memory locations A and B, add them together, and store the result in a memory location addressed as C.

4. The fourth statement causes the computer to print the contents of the memory location addressed as C on the specified output device. If the output device is not specified, then it would be printed on the default output device, which is usually the computer screen.

5. The fifth statement causes the computer to display a message on the computer screen asking if the user wants to perform more additions. The actual message to be displayed needs to be included in the program. Of course, we can include it in the algorithm also.

6. The sixth statement causes the computer to wait for the user to enter an answer and receive it. The answer can be "Yes" or "No." We may specify that his/her answer be stored in a memory location or directly be taken to the CPU. We need to specify the alternative in the program.

7. The seventh statement causes the computer to make a decision based on the answer supplied by the user.

8. Once the decision is made, the computer begins executing the program all over once again if the user-supplied answer was "Yes."

9. If the answer was "No," the computer would stop running the program. This would include the following steps:

   a. Remove the program from the list of programs under execution.

   b. Release the RAM that was allocated for storing the program and the data items used by the program.

   c. If any data files were opened for use by the program, they would be closed.

   d. It would release the input and output devices used by the program for use by others.

In this manner, we can develop other algorithms. What should be the level of granularity that is desirable? There is no fixed answer to this question. The level of detail and granularity of the algorithm depends on the next person, the programmer who needs to understand the algorithm and convert it into a computer program using a programming language. When the programmer is a reasonable expert and has the adequate experience in computer programming, we may not need to include excessive detail and keep the granularity at a coarse level. If the programmer is a trainee, then we need to include greater detail and finer granularity in the algorithm.

But how do we develop an algorithm?

Developing Algorithms

The following are the steps in developing an algorithm:

1. Study the problem at hand. It can be a mathematical problem, or it can be a transaction problem that we are attempting to develop an algorithm to solve.

2. Break the problem down into its component parts.

3. Write down the steps in arriving at the solution for each of the components of the problem at hand.

4. Further subdivide each step into multiple sub-steps as necessary. In this step, we need to divide in such a manner that you understand the procedure to arrive at the solution.

5. Review the algorithm and assess if a third person can understand the steps in arriving at the solution for the problem, beginning at the first step and traversing the steps in their sequential order.

6. You can depict the solution graphically using a flowchart and analyze it for accuracy.

7. Revise them to ensure that a third person can understand the steps without any difficulty.

8. It will always be better to get a peer (a person working in a similar capacity and carrying out similar activities as you are) to review the algorithm and implement the feedback.

Now the algorithm is ready.

If you are familiar with algebra at least to high school level, you would be familiar with algorithms, as algebra uses algorithms to solve problems. If you are not familiar with algebra, then you need to develop the skill in developing algorithms by mastering the skills and following the earlier guidelines rigorously. It would help greatly if you are an understudy with an experienced person.

I referred to the word "transaction" earlier. By "transaction," I meant a business transaction. A business transaction has the following attributes:

1. It involves an exchange of things. A sales transaction involves exchange of goods from one party and money from the other party. In a ticket-booking transaction, the ticket is exchanged for money. In a material-issue transaction in a warehouse, a material requisition is exchanged for material. In a registration transaction, information is exchanged for confirmation of registration. In a cash-payment transaction, cash is exchanged for a receipt.

2. It involves information, which can be a single data item or multiple data items that need to be stored for future reference.

3. It has a set of rules to be adhered to in order to successfully conclude the transaction. These rules may include exception-handling rules, too.

4. It involves certain checks for accuracy to ensure that the transaction is properly concluded.

5. In some cases, the transaction may not be successfully concluded, but it may need to be recorded all the same.

In such scenarios, to develop a suitable algorithm to develop a program, we need to study the workflow of the transaction. The workflow includes the steps in carrying out the transaction, the rules to be adhered to, the inputs, and the outputs. Then follow the steps outlined earlier to develop the algorithm.

Of course, there are other scenarios for developing computer programs including development of software for hardware/machine control, real-time software, system software, and so on. To develop algorithms for such programs, we need to understand the hardware involved and its functioning. Even that follows essentially the same steps in developing algorithms as business software.

Now let us look at flowcharts that I mentioned earlier to evaluate the algorithms.

Flowcharts

Flowcharts, as the name implies, are charts that depict, in this case, the flow of the execution of instructions by the computers. Alternately, flowcharts graphically depict our intention of the chronological steps to be executed sequentially by the computer to arrive at the desired solution for the problem. These are graphical methods and assist us in visualizing the process more easily. It is said that a picture is worth a thousand words. Flowcharts were the only tool for computer programmers until very recently for developing algorithms for computer programming, but lately, other graphical tools have been developed, which include Data Flow Diagrams (DFDs), class diagrams, unified modeling language (UML), structure charts, and so on. But still, at a program level, a flowchart is the best technique to develop an appropriate algorithm and to evaluate its accuracy in arriving at the desired solution flawlessly.

Should we use both algorithms and flowcharts? It depends on the complexity of the problem at hand. If the problem is simple in nature, we may use either algorithm or flowchart. If the problem is complex in nature, then it may be better to use both the algorithm and the flowchart. One point may be noted here, and that is that a flowchart lends itself easier to read, analyze, and interpret than an algorithm, especially when the problem is complex in nature.

Flowcharts use the following symbols:

- ⬭ Depicts the beginning or the ending of the flowchart. It is either the first block in the chart or the last one in the chart.

- ☐ Depicts a step in the process. It depicts the action to be performed.

- ◇ A decision block. It handles a total of three possible outcomes for a decision scenario.

- ▱ Depicts data either as input or output. We use this block when we have not decided which device to use for input or output.

- ⬓ Depicts paper-based output or a device that handles paper. Scanners are input devices handling paper, and printers are output devices that handle paper.

- ⬯ Depicts data from or to a disk drive.

- ⬭ Depicts a computer screen.

- ⌐ Depicts the connector between blocks. The direction of the arrow depicts the flow of execution.

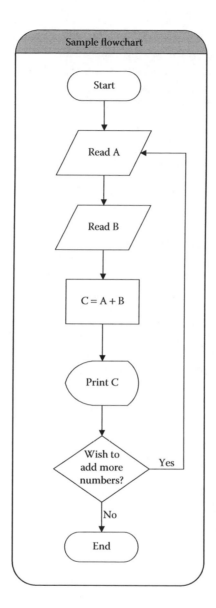

FIGURE 5.1
Flowchart to add two numbers.

There are some more symbols that are used in preparing flowcharts. You can learn them as you get more and more proficient in flowcharting. The earlier symbols suffice to begin flowcharting.

Let us depict the algorithm we enumerated above to add a couple of numbers. It is depicted in Figure 5.1. Of course, we can extend it to add more numbers. We can also combine all read operations into one process step.

From this, we can enumerate the general rules followed while preparing the flowcharts.

1. The first and last blocks are horizontal ellipses. All other blocks are embedded between these two blocks. It is the normal practice to caption the first block as "Start" and the last block as "End" or "Stop."

2. The desired flow of execution is depicted using a connector line with an arrow-head at one end.

3. The flow is generally forward, that is, from top toward the bottom and from left toward right. An exception to this rule is when the flow goes back as a result of a decision as shown in the Figure 5.1 at the decision block.

4. We can have any number of blocks between the Start block and the End block.

5. When the flowchart cannot be accommodated on one single sheet, we can continue it on to another sheet by using a connector. A connector is a simple circle with a number inside it. The number on the connector at the end of the sheet would be the same as the number inside the connector at the beginning of the next sheet where the flowchart is continued.

I will end the discussion on the algorithms and flowcharts by developing the algorithm and its flowchart for determining if the given number is a prime number or not. What is a prime number? It is a whole number that is divisible only by itself and 1, leaving the remainder as zero. Any number can be divided by 1 leaving a remainder as zero. So, a prime number is a positive number that cannot be divided by any other number except by itself with zero as the remainder.

In mathematics, we divide the number with the first divisor as 2 and then incrementing it by 1 until the divisor just crossed half the value of the number. This can be illustrated by an example:

Let us take a number 11 and find if it is a prime number

1. First, we divide it by 2 and see if the remainder is zero.

2. Then we divide it by 3 and see if the remainder is zero.

3. Then we divide it by 4 and see if the remainder is zero.

4. Then we divide it by 5 and see if the remainder is zero.

5. Then we divide it by 6 and see if the remainder is zero.

6. As the divisor is more than half of 11, we know that there is no point in further incrementing the divisor and performing the operation once again. Now, so far, the remainder happened to be a non-zero. So, we determine that 11 is a prime number and stop further effort.

Now, we can generalize the previous steps and write the algorithm as follows:

1. Read the number, M

2. $N = 2$

3. Divide M by N

4. If the remainder is not equal to zero, then go to 6

5. If the remainder is equal to zero, then go to 9

6. If $N > M/2$ then go to 11

7. $N = N + 1$

8. Go to 3

9. Print "M is not a Prime Number"

10. Go to 12

11. Print "M is a Prime Number"

12. End

Now, this algorithm is depicted pictorially as a flowchart in Figure 5.2. Of course, we can modify this flowchart. We can also have a different algorithm to arrive at the solution. I have used a simpler version to introduce you to the techniques of developing algorithms and flowcharting.

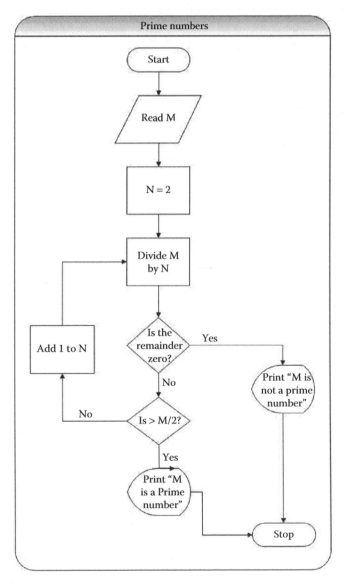

FIGURE 5.2
Flowchart to determine prime numbers.

Handling Data in Real-Life Programs

In the earlier algorithm, we neglected the aspect of handling data and assumed that the data is input using the default input device. But in real life, the data can be voluminous and it may not be practical to input all the data from the keyboard. All data comes from secondary storage and the most common storage medium is the magnetic disks. Most of the data comes in from the databases. We discussed the data files and databases in Chapter 3. The methods to access data from data files and databases will be dealt with in the coming chapters.

6

Statements and Assignment Statements

Introduction

As we have noted in the earlier chapters, computer programs are a sequence of instructions to computer giving details of what needs to be accomplished and the details of the data to be used. Instructions to the computer are given using statements. "Statement" is a very frequently used word in general parlance. It is similar in meaning to information, but it connotes authentication and affirmation and is expected to be much more reliable. Statements are made in courts of law, by politicians, and in press conferences by others. In general, a statement has the following attributes:

1. It covers some aspect comprehensively.
2. It is accurate in syntax and semantics.
3. It is issued by a person concerned with and knowledgeable of the situation at hand.
4. It is mostly written and is on paper.
5. When it is issued orally, it is authenticated and authorized.
6. Statements are not given frivolously.

That is why the word "Statement" is used in respect of computer programs. In the earlier days, a statement was accommodated in one line. A line and a statement were synonymous. In those days, a line could accommodate only 80 characters, but a need arose to span the statement across multiple lines and computer programmers came up with a method to continue the statement onto the next line. But modern computer screens can accommodate more than 80 characters per line, and some screens can accommodate up to 255 characters per line. Others have the facility of horizontal scrolling and thus accommodate more than 80 characters per line. Most modern programming languages permit up to 255 characters per line and also allow multiple lines in a single statement. Presently, a statement in a computer program has the following attributes:

1. A statement contains at least one computer instruction.
2. A statement can contain multiple computer instructions.
3. It contains both instructions and the references to the data items that are processed by the instruction.

4. A statement can be contained in one line, or it can span across multiple lines. When the statement spans across multiple lines, it is the general practice to prefix the continuing statement with a specially designated character. The continuation character differs from one programming language to another. In the modern programming languages, this is not a requirement.

5. Generally, a statement is terminated by a specially designated character. The most commonly used statement terminator character is the semicolon (;).

6. Statements must follow the prescribed syntax.

All programming languages specify a set of syntax rules for writing statements in that language. Let us discuss the syntax here.

Syntax

Each statement in a computer program has to adhere to a set of rules specified by the programming language. This set of rules is commonly referred to as syntax for that language. Syntax is defined as "the arrangement of words and phrases to create well-formed sentences in a language" by Google. Merriam Webster's Dictionary defines syntax as "arrangement of words in a sentence, clauses and phrases." Wikipedia defines syntax as "the set of rules, that defines the combination of symbols that are considered to be correctly structured document or fragment in that language" in the context of computer programming.

Syntax for programming languages has the following characteristics:

1. The program begins with a specially designated keyword to indicate the beginning of a program.

2. The program ends with a specially designated keyword that tells the computer that the program execution is completed and necessary closing actions can be taken.

3. All program statements have to be embedded between these program-beginning and ending keywords.

4. It uses words and phrases defined by the programming language. The words are of two types: words that are defined by the programming languages and the words that are defined by the programmer. The words defined by the programming language are often referred to as "keywords."

5. Keywords include certain symbols, such as arithmetic operators (+, −, *, /, ^), parenthesis, relational symbols (<, >, =), and any other programming-language-specific symbols.

6. The keywords and the words defined by the programmer are mutually exclusive. The keywords cannot be used by the programmer for any purpose other than the purpose defined by the programming language. User-defined words are usually data declarations.

7. Generally, the statement begins with a keyword and is followed by user-defined words and, in some cases, other keywords. Exceptions to this rule are found especially in arithmetic statements, in which case the statement can begin with a programmer-defined keyword.

8. There are rules for arranging the keywords, symbols, and the programmer-defined words, and these must be adhered to strictly. No deviation is allowed from these rules under any circumstances. Of course, some programming languages like COBOL (Common Business Oriented Language) allow the use of some words for documentation purpose.

9. A statement in most programming languages is terminated by a specially designated character. A semicolon (;) is the most commonly used statement termination character.

Non-adherence to these rules of syntax would give rise to what are referred to as "syntax errors." The first step in converting the source code to executable code is checking the program for syntax errors. The compiler begins the conversion of source code to executable code only when there are absolutely no syntax errors in the source code of the program. When the program compiler encounters a syntax error, it halts all further steps to convert the source code to executable code and outputs the syntax errors so that the programmer can correct them and resubmit the program for compilation.

Types of Statements

The statements are classified in two ways: by the function they perform and by the complexity of the statements. By the function performed, statements are classified into the following types:

1. *Declaration statements*: Declaration statements are used for declaring the data being used in the program. Declaration statements also declare any custom or special libraries other than those that were supplied by the vendor in the SDK (Software Development Kit) used by the program.

2. *Assignment statements*: Assignment statements assign a value to a variable already declared earlier in the program. The value to be assigned may be another variable, a constant value supplied within the program, or is the result of evaluation of an arithmetic expression. We will learn about expressions in the later chapters.

3. *Input/output statements*: These statements, as the name itself implies, receive data from the outside world or deliver data to the outside world.

4. *Control statements*: Control statements halt the program execution from going to the next sequential statement and can shift the execution to another statement, which need not be the next statement in the sequential order. They assist us in making programmed decisions and processing the information as necessary.

5. *Loops*: Loops are blocks of statements that are executed a number of times repeatedly based on a condition. These are very useful statements in programming.

6. *Error-handling statements*: Even when we take extreme care to avoid errors when the program is running, still errors can creep in due to errors in the data, unexpected results, and the environment. We use error-handling statements to trap the error conditions and take corrective actions to ensure that the program closes smoothly or to move the program away from the error condition to a normal condition.

7. *System calls*: Usually, programs are written using the facilities provided by the programming language. Sometimes, we may need to utilize the facilities provided by the operating system that are not provided by the programming language. This is achieved by the system call statements. These are utilized more by advanced programmers.

8. *Inter-process communication*: These statements are also not the run-of-the-mill statements utilized by every programmer. These are utilized by advanced programmers. These statements are utilized when two programs running in the computer concurrently need to communicate with each other for an exchange of information. These are heavily used in the development of the system software and real-time software.

9. *Interrupt handling*: These statements are used by programmers in the development of software for device handling. Any device connected to the CPU places an interrupt on the CPU when it needs to communicate. These are also advanced statements used in the development of system software and real-time software.

10. *Device handling*: All the peripherals including printers, scanners, other machines, airplanes, and rockets need special software. These are usually referred to as device drivers. Device drivers are software programs that handle the devices and make them perform the desired functions. These have special calls and syntax.

11. *Starting and ending statements*: Every program begins with a special statement. This tells the computer that the program begins execution, and this instruction is followed by a series of instructions to be executed by the computer. Similarly, every program ends with a special statement. It tells the computer that the execution of the program is completed without a hitch so that it can close the program, delete it from the RAM, and release all the RAM allocated for the data items used in the program, as well as release the devices used by the program. These are part of every program.

12. *Documentation statements*: One thing is certain in software development and programming. Every developed program put into production needs to be modified and enhanced during its lifetime. The original programmer who wrote it may have departed or is not available currently to maintain the program. Therefore, the programs need to be written in such a manner that another programmer would understand and be able to maintain it. The programming languages are normally cryptic in nature. So, all programming languages provide a feature called as "commenting statements." The commenting statements are completely ignored by the computers while compiling and preparing the executable code. These are for the reference by the programmers maintaining the programs. Usually, a specially designated character is prefixed to the statement to indicate that this statement is a commenting statement used for the purpose of documenting the program logic.

By complexity of the statements, they are classified into:

1. *Single statements*: Single statements perform one function and are usually contained in one line or a set of continuing lines. Assignment statements are usually single statements.

2. *Compound statements*: Compound statements have multiple statements and perform more than one function and span across multiple independent lines. These are also referred to as "block statements," or as a block of statements. Control statements and loops are usually compound statements.

In this chapter, we will discuss in detail the assignment statements. We will discuss the documentation statements, starting and ending statements, declaration statements, system calls, inter-program communication, interrupt handling, and device handling statements in a separate chapter. I have dedicated a separate chapter for each of the other statements.

Assignment Statements

Assignment statements form the major chunk of any computer program. It is the statement in which actual work is carried out. Assignment statements are taken from the mathematics field. In mathematics, an assignment statement looks like this:

$$a \leftarrow b + c + d - e$$

The import of this expression is:

1. Evaluate the expression of $(b + c + d - e)$.
2. Assign the resulting value to "a."
3. The arrow symbol is the symbol of assignment. The arrow head indicates the direction of assignment.
4. Generally, the arrow head points to the left side of the expression.
5. Only one variable is allowed on the left side of the arrow.

The same rules apply to assignment statements in computer programming too, but the assignment symbol is not the arrow sign. In most programming languages, it is the equal to (=) sign. An assignment statement in a computer program looks like this:

$$A = B + C$$

The import of this expression is:

1. Take the contents of memory location B
2. Take the contents of memory location C
3. Add B and C
4. Assign the result of the addition to memory location A

One deviation that assignments in computer programming have over the mathematical assignment statements is that the variable on the left-hand side of the assignment symbol can also be there on the right-hand side of the assignment symbol. For example, "$A = A + B$" is permitted in computer programs but not in mathematical assignments. This statement results in:

1. Take the contents of memory location A
2. Take the contents of memory location B

3. Add A and B

4. Assign the result to memory location A

In fact, this type of statement is very common in computer programming. Generalizing the assignment statements, we can enumerate the following rules:

1. The assignment statements would contain the assignment symbol, occurring only once, which is usually the equal to (=) sign.

2. On the left-hand side of the assignment statement must be a single variable.

3. On the right-hand side of the assignment symbol, there can be:

   a. A constant, which can be a number or a literal (a string of characters enclosed in quotation marks).

   b. A mathematical expression containing all variables, all constants, a combination of variables and constants, or a combination of expressions.

   c. The variable on the left-hand side of the assignment statement can be part of the expression on the right-hand side of the assignment statement.

4. During execution of the program, the expression on the right-hand side of the assignment symbol is evaluated and the result would be assigned and stored in the memory location symbolized by the variable on the left-hand side of the assignment symbol.

We will learn in detail about the expressions used in computer programming in later chapters. Let us first learn about initialization statements. Initialization statements are a type of assignment statements in which an appropriate value is assigned to a variable before it is used in an expression. Here are a few examples of initialization statements:

```
Count   = 0;
Salary = 0
Totals = 0
Names   = "" (that is a set of blank spaces)
Ismarried = TRUE
```

Statements similar to the earlier examples are frequently found in computer programs in which we assign a constant, usually, zero to a variable. These are referred to as "*initialization statements.*" When we declare a variable, the compiler assigns a memory location to it. What default value would be assigned to that variable by the computer when the program execution begins? Some powerful computers and programming languages assign an appropriate value such as zero for numeric variables and blank spaces for character variables. But most computers and programming languages assign a special value referred to as "NULL" to the variables immediately upon the declaration. This value is not usable in any expressions except in comparisons. When we use the variable with NULL value in expressions and assignment statements, the computer throws up an error that may result in abrupt termination of the program execution. Therefore, to avoid such uncontrolled errors, programmers must assign an appropriate value to the variables immediately after declaring them. This process is referred to as "initialization."

While zero is the most assigned value for numeric variables, and blank spaces for character variables, in initialization statements, other values are also assigned to the variables. We may also need to initialize the variables multiple times in the program depending on the specific condition in the program. Most errors in getting wrong results is due to forgetting to initialize the variables at appropriate locations within the programs.

Here are some of the examples of other assignment statements we can find in computer programs:

```
Wages = No_hrs_worked * Hourly_rate
Income_Tax = Total_Earnings * 0.25
Salary = bpay + (bpay * 0.1) + (bpay * 0.2) - (bpay * 0.15)
Sales_Commission = Monthly_Sales * 0.05
Isshemarried = FALSE
Whatsthename = "Harold"
```

As you can see, Wages, No_hrs_worked, Hourly_rate, Income_tax, Total_Earnings, Salary, bpay, Sales_commission, Monthly_Sales, Isshemarried, and Whatsthename are variables. As you can also see, on the left-hand side of the assignment symbol, there is only one variable and on the right-hand side of the assignment symbol, there is an expression.

Utility of Assignment Statements

Assignment statements are used for the following purposes:

1. We use assignment statements for initializing the variables used in the program.

2. We use assignment statements to assign values to the environment and other variables, especially the configuration information of the user environment. For example, we often select screen colors to suit our individual tastes in many applications. When using applications, we set our preferences to a host of settings made available for us by the software package. Most computer operating systems make many options for the individual users to choose from. We set our user ID and password and keep changing the password often. These are achieved by assigning our preferences using assignment statements.

3. We use assignment statements for evaluating mathematical expressions and arriving at the solution to a mathematical problem. Computers can solve simple/complex arithmetic problems used in business transactions, including computation of salaries for employees, the prices in sales transactions, costs in various aspects of management, decision support systems, and interest computations in banking, as well as solving complex problems like calculus, linear programming, transportation problems, matrix algebra, and, for that matter, any mathematical problem for which an algorithm is available.

4. We use assignment statements for concatenating strings to form a new string of characters. Why should we concatenate strings of characters? There are many practical uses for such an operation. One such use that comes to mind readily is to form a connection string used to open databases. We usually store the first name, middle name, and the last names of individuals in our application but we output the complete name, and this necessitates concatenating the three together.

5. We often use assignment statements in writing information to data files and database tables using assignment operations.

Best Practices in Writing Assignment Statements

It is always better to write short expressions on the right-hand side of the assignment operator. It is difficult to understand and maintain long expressions during debugging or program maintenance.

When it becomes essential to write long assignment statements, it is better to break the statement into multiple statements and write one below the other rather than write a very long assignment statement.

I advocate limiting the statement to a length where it is not necessary to horizontally scroll the screen to read the statement. Scrolling horizontally renders the program reading and understanding very difficult.

7

Arithmetic, Relational, and Logical Expressions

Introduction to Expressions

As we have noted in Chapter 6, expressions are used in statements. Almost every statement contains at least one expression. Merriam Webster's dictionary defines an expression generally as "an act of expressing" and "an act of making your thoughts, feelings etc. known by speech, writing or some other method," and in the context of computer programming, thus, "a mathematical or logical symbol or a meaningful combination of symbols". Wikipedia defines an expression thus: "an expression in computer programming is a combination of explicit values, constants, variables, operators and functions, that are interpreted according to the particular rules of precedence and of association for the specific programming language."

Let us enumerate the attributes of expressions in the context of computer programming to understand them better:

1. An expression may consist of at least two variables, a constant, or a combination of variables and constants.
2. The variables and constants in the expression are combined using mathematical, relational, or logical symbols, generally referred to as "operators."
3. The expression is amenable to evaluation using arithmetic, relational, or logical rules.
4. The evaluation of an expression in a statement yields only one value that can be used for assignment to a variable or in programmed decision making. That is, expressions can be used in assignment statements and control statements. Expressions are not used in a stand-alone mode.
5. Expressions, when used in assignment statements, must be on the right-hand side of the assignment symbol. An expression can never be on the left-hand side of an assignment symbol.

Types of Expressions

Expressions used in computer programming are of three types, namely:

1. Arithmetic expressions
2. Relational expressions
3. Logical expressions

Let us discuss each of these types in greater detail in the following.

Arithmetic Expressions

Arithmetic expressions are used to solve arithmetic equations and compute the required values. These expressions can include both variables and constants. The variables and constants used in arithmetic expressions have to be numeric in type. Of course, some programming languages allow use of character type variables and constants in arithmetic expressions. But, they can be used only for addition. When two are more strings of characters are used in an arithmetic expression using an addition operator, they would be concatenated together. The variables and constants in the arithmetic expressions are joined together to form an expression by the arithmetic operators. The following are the arithmetic operators:

1. Addition symbol   +
2. Subtraction symbol   − (a dash)
3. Multiplication symbol   * (an asterisk)
4. Division symbol   / (a slash)
5. Exponentiation symbol   ^ (a caret)
6. Open parenthesis   "("
7. Close parenthesis   ")"
8. For square root, there is no symbol allocated. It is usually achieved by a library routine. The specific library routine used for finding the square root differs among programming languages. Usually, it is either "SQRT" or "SQR."

In real-life mathematics, we use curly braces {and} and square braces [and] along with curvy braces, "(" and ")", when more than one set of parentheses is needed in the mathematical expression. In computer programming, we use only the curvy braces in arithmetic expressions. Here are a few valid examples of arithmetic expressions:

- a + b
- a + b − c
- a + b * c
- a + (b * c)
- (a + b) * c
- a − b * c / d
- (a + b) ^ 2
- (a + b) ^ 2 − (a − b) ^ 2 / d
- ((a + b) ^ 2 − (a − b) ^ 2) / d

Precedence Rules of Evaluation

While evaluating arithmetic expressions, computers usually follow these precedence rules:

1. The evaluation proceeds from left to right when the precedence level of the operators is same.
2. The addition symbol and the subtraction symbols have the lowest priority in the precedence of evaluation. They both have the same precedence.

3. The multiplication and the division symbol have the same precedence. They have higher precedence than the addition and subtraction symbols.

4. The exponentiation symbol has higher precedence over the multiplication and division symbols.

5. Parentheses have the highest precedence than the rest of the arithmetic operators.

Now using these rules, let us see how the computer evaluates the previous example expressions:

1. a + b—there is only one operator. The computer just adds the values of both the variables and delivers the result.

2. a + b − c—here we have two operators with the same precedence. Therefore, the computer will add the value of a with the value of b and then it would subtract the value of c from the sum and deliver the result.

3. a + b * c—here we have two operators with different precedence levels. Therefore, the computer will first evaluate the part of the expression joined by the operator with the higher precedence. So, it will multiply the value of b with the value of c. Then it will add the result to the value of a and then deliver the result.

4. a + (b − c)—here we have three operators with different precedence levels. The computer will evaluate the expression within the parentheses. So, the value of c would be subtracted from the value of b and then the result would be added to the value of a. Then the result would be delivered.

5. (a − b) * c—here we have three operators with different precedence levels. The computer will evaluate the expression within the parentheses. So, the value of b would be subtracted from the value of a and then the result would be multiplied by the value of c. Then the result would be delivered.

6. a − b * c / d—here we have multiple operators with different precedence levels.
   a. Following rule #1 of evaluation, the evaluation proceeds from left to right. Therefore, b would be multiplied by c.
   b. The result would then be divided by d.
   c. Then the result would be subtracted from the value of a.
   d. Then the result would be delivered.

7. (a + b) ^ 2—the evaluation in this expression is rather straightforward. The expression in the parentheses would be evaluated first. Then it would be raised by the value of the exponent, 2, and then the result would be delivered.

8. (a + b) ^ 2 − (a − b) ^ 2 / d—here we have multiple operators with different precedence levels.
   a. First, the computer would evaluate the expressions in the parentheses.
   b. It would then divide the result of the expression (a − b) ^ 2 by d.
   c. Then the result would be added to the result of expression (a + b) ^ 2.
   d. The result would be delivered.

9. ((a + b) ^ 2 − (a − b) ^ 2) / d—here we have multiple operators with different precedence levels.

   a. First the computer would evaluate the expressions in the inner-most parentheses. Then the outer parenthesis would be evaluated.

   b. Then the result of expression $(a - b)\,^\wedge\,2$ would be subtracted from the result of
expression $(a + b)\,^\wedge\,2$.

   c. Then it will divide the result by d.

   d. Then the result would be delivered.

As you can see, the expressions in bullet points #8 and #9 are similar except the placement of parentheses, which changes the way the expression is evaluated. Both expressions
would deliver different results.

   Therefore, we need to be careful when programming the arithmetic expressions, especially when using the operators with different precedence levels. As a precaution, it is recommended to use the parenthesis operators liberally to avoid confusion about the precedence
values to obtain the desired and accurate result. It is also recommended to avoid writing long
arithmetic expressions, as it can be very confusing when debugging the program. It is better
to break long expressions into multiple smaller expressions and write them in different lines.

   When using arithmetic expressions in arithmetic statements, the following precautions
are suggested to obtain an error-free result:

1. The variable receiving the result of the expression needs to be able to accommodate the result. If we are using floating-point numbers of the single-precision type,
we need to have a variable of double-precision type on the left-hand side of the
assignment symbol.

2. If we assign the value to a variable that has a lesser capacity than the result of the
arithmetic expression, the result would be truncated to the size of the variable.
This results in an erroneous result.

3. If we use all integer type values in the arithmetic expression, it is better to assign
the result to a variable of single-precision type.

4. If we use all single-precision or a combination of integers and single-precision
variables in the arithmetic expression, then we better assign the result to a variable
of double-precision type.

Important: One major precaution we need to take while defining the arithmetic expressions
is to ensure that the denominator never becomes zero. Anything divided by zero results
in infinity. But we can never know what the real-life data will throw at our program. So,
whenever we need to use a division operator in our arithmetic expression, it is essential
to check if the denominator is zero or NULL before evaluating the expression. This, in
addition to forgetting the initialization statements, is a major reason for abrupt abortion of
the program execution causing damage to data in files and delivering erroneous results.
Multiplication by zero is alright, as the result would be zero, but division by zero is unac-
ceptable. So, we must take care to ensure that the denominator never becomes zero while
writing programs using arithmetic expressions.

Best Practices in Forming Arithmetic Expressions

1. Use parentheses liberally while forming arithmetic expressions, especially when the expression is too long. Which expression is too long? While there is no universal agreement, I would say, based on my experience, that any expression that has a combination of addition, subtraction, multiplication/division, and exponentiation is too long. Whenever we combine multiplication, division, and exponentiation with addition of subtraction operators, we are making room for confusion. As a rule, whenever we combine operators of different precedence levels, it is better to use parentheses to remove any room for doubt and confusion.

2. Limit the length of the expression to fit the screen without the necessity for horizontal scrolling. The number of characters a computer screen can accommodate depends on the screen resolution. So, make a decision about the number of characters in an expression depending on the screen resolution.

3. Even if the screen can accommodate a greater number of characters in a line, it is better to divide the expression into multiple lines and statements for the sake of clarity. I would advocate to limit the expression in one line to about 60 characters.

4. When using parenthesis, the process of evaluation becomes obscure when we use too many of them. As a rule, I would advocate limiting the opening braces to a maximum of three at one place. That is, if the first character begins after placing three opening braces, it is alright, but if you have to type in four opening braces before beginning the first character, it is not suggested. If you are a serious programmer, at one time or the other, you will be counting the number of opening braces and the number of closing braces and then match the numbers. It may be hard to believe, but we programmers of mathematical software spend a considerable amount of time ensuring the number of opening braces is equal to the number of closing braces. This is also a major cause in giving rise to syntax errors and then logical errors. Placement of braces at the wrong places is also a reason for giving inaccurate results.

5. When we use arithmetic expressions in assignment statements, we need to ensure that the variable designated to receive the result is of adequate size to accommodate the result. When the receiving variable does not have adequate capacity, then the result would be truncated to be accommodated in the variable. While the expression is evaluated correctly, the delivered result would be inaccurate.

6. Whenever a division operator is used in the expression, we need to ensure that the denominator does not become zero. So, before programming an arithmetic expression, we need to check and ensure that the denominator is not zero. We can do this using control statements and relational expressions. I would suggest programming the division operation on a separate line by itself.

Relational Expressions

Relational expressions relate one expression with another and determine if the relationship between the two is true or false. Relational expressions are used for decision making and are used in control statements. They are not used in assignment statements. The expressions in the relational expressions are joined together by the relational operators. Relational operators specify the type of relation to be established between both the expressions. The relational operators are given in Table 7.1.

While parentheses are not relational operators, parentheses are often used to enclose a relational expression. It is a good practice of programming to enclose the entire relational expression in a set of parentheses. The relational operators given in Table 7.1 are the popular representation used by most programming languages. But variations can be seen in some programming languages. For example, COBOL uses full text, such as LESS THAN, GREATER THAN, NOT EQUAL, and so on. In some programming languages, the characters le, ge, eq, and so on, are used. You need to check the programming language manual or help pages to know exactly what relational operators are used in that language.

As noted earlier in this chapter, an expression can contain a variable, or a constant, or a combination of variables and constants. We need to follow the following rules when writing relational expressions:

1. The expressions need to be on different sides of a relational operator. The relational operator would be in the middle with one expression on its left-hand side and the other expression would be on its right-hand side.

2. A relational expression can contain only one relational operator. We cannot have more than one relational operator in a relational expression.

3. Both the expressions included in the relational expression need to be of the same data type. If the expression on the left-hand side is of numeric type, then the one on the right-hand side needs to be of numeric type. If one of the expressions is of numeric type, the other cannot be of any type other than numeric.

4. If both expressions are of numeric type, their precision could, however, be different. If one is of integer type, the other can be of single- or double-precision type. The result, however, would depend on the actual value and the relational operator.

5. Some programming languages allow different data types of expressions. One can be of numeric type and the other can be of character type. But the results would be unpredictable. That is, the programming language does not take the responsibility

TABLE 7.1

Relational Operators

Relation	Symbol	Characters
Equal to	==	eq
Not equal to	!= or <> or /=	ne
Greater than	>	gt
Greater than equal to	>=	ge
Less than	<	lt
Less than or equal to	<=	le

of finding logical programming errors. It is incumbent on the programmer to write defect-free programs. Even if the programming language permits, it is not right to write relational expressions using two expression of differing data types on either side of the relational operator.

6. The result delivered by the evaluation of a relational expression is either TRUE (one or YES) or FALSE (zero or NO).

We need to take note of one important aspect here. While it is permitted to use constants in relational expressions, it is pointless to use only constants on both sides of the relational operator. What is the point in comparing 3 and 4 to see if 4 is greater than 3?

All relational operators have lesser precedence than all arithmetic operators. So, if an arithmetic expression is included in a relational expression, it would be evaluated first before evaluating the relational expression. Now let us consider some examples of relational expressions:

1. x > y—this expression compares value of x with the value of y. Then, if the value of x is more than y, it will return the result of TRUE. If the value of y is either equal to or is more than the value of x, then it will return a result of FALSE.

2. a + b > c—in this expression, it will add the value of a to the value of b and only then it will compare the resultant value with the value of c. It will finally return a result of TRUE only when a + b happens to be more than the value of c.

3. (a + b) > c—this expression is similar to the expression in bullet #2 except that a parenthesis is used to make the process of evaluation more explicit.

4. (a + b) < (c + d)—this has arithmetic expressions on both sides of the relational operator and both are enclosed in parentheses. Both these expressions would be evaluated first and then the results would be compared. Then the appropriate result, be it TRUE or FALSE, would be delivered.

5. ((a + b)/c) >= (x + y/z)—this is another example of two arithmetic expressions being compared. In the expression on the left-hand side of the relational operator, the value of a would be added to the value of b and then the sum would be divided by c. In the other expression, the value of y would be divided by the value of z and then the quotient would be added to the value of x. Then the results of both the expressions would be compared to deliver the result of the relational expression.

Now, we need to understand the difference between some confusing symbols. Of course, the meaning of symbols is not confusing really, but if we are not careful, we can use them wrongly. Let us consider some examples. Let us assume that the value of a is 4; value of b is 4; value of c is 4.001. Now let us use these values in our examples:

1. (a == b)—this would evaluate to TRUE, as the value of both a and b is 4.

2. (a != b)—this expression evaluates to FALSE, as the values of both a and b are equal.

3. (a <= b)—this expression would evaluate to TRUE. B is not less than a, but it is equal to a.

4. (a >= b)—this expression would also evaluate to TRUE. A is not greater than b, but it is equal to b.

5. (a == c)—this evaluate to FALSE because, while the difference between the values of a and c is very small, it is still significant. But a and c are not equal, however small the difference may be. We human beings may consider the difference to be insignificant for all practical purposes, but a computer would consider any difference, irrespective of its magnitude, to be significant.

6. (a != c)—this expression evaluates to TRUE, as the values of a and c are not equal.

7. (a >= c)—this expression evaluates to FALSE. The value of c is certainly greater than that of a in the view of the computer.

8. (a <= c)—this expression evaluates to TRUE as a is certainly less than the value of c.

What we have to note here is that for writing good programs, we need to ensure that both variables being compared are of same precision. If the variable on the left-hand side is integer, the right-hand side variable also needs to be integer to ensure accurate results. If we compare an integer value with a single- or double-precision value, the result is rather unpredictable. So is the case when we compare a double-precision variable with an integer or a single-precision variable; the result is unpredictable. If we wish to write reliable programs that deliver consistent results, we need to ensure that the precision is the same in numeric variables used in the relational expressions.

Now let us consider some character data. Let us assume that the variable fname contains a value of "Thomas;" the variable lname contains a value of "thomas;" the variable mname contains a value of "Tho mas;" the variable x contains a value of "THOMAS;" and the variable y contains a value of "thomas." Now let us see some examples:

1. (fname == lname)—this expression evaluates to FALSE. While both contain the same spelling, the first letter in variable lname is lowercase. This is significant for computers, even though we human beings consider the difference as insignificant.

2. (fname == mname)—this expression also evaluates to FALSE. There is a space between "o" and "m" and this is considered as a significant difference by the computer. Also, variable fname contains six characters while the variable mname contains seven characters.

3. (x == y)—this expression too evaluates to FALSE. While the value of x contains all uppercase letters the value of y contains all lowercase letters. This difference is significant in the view of computer.

Therefore, when we compare two-character strings, we convert both variables into either lowercase or uppercase before comparing them to deliver results that are sensible to human beings.

We human beings are very inconsistent when we come to spelling our names and use different spellings for the same name. Some of this has to do with the cultural, national, and linguistic backgrounds we came from. So, locating a certain record based on imperfect information is difficult with the aforementioned relational expressions. Instead of trying to locate the required string of characters using exact match, we like to locate the record by using a part of the total string to locate a set of records from which we can select the required one manually. Sometimes, we may locate a set of records as coming from the same group for some purpose. What we need is not the exact match of the string but finding if a longer string of characters contains within it a shorter string of characters. For example, we may need to locate records whose name is either Katherine or Catherine.

We know both names are pronounced same but spelled differently. Instead of searching twice, we wish to locate all records in whose name, the string "rine" is embedded. None of the aforementioned relational operators would accomplish our need.

This is a well-recognized need. This is achieved by a pre-defined library function in some programming languages. Some languages provide a relational operator. Generally, the dollar symbol "$" or "$$" is used as this relational operator that permits searching a string within another longer string. Let us assume the variable fname contains the value "Katherine;" the variable gname contains the value "Catherine;" and the variable hname contains the value of "Kathereen." Let us also assume that a variable xyz contains the value "Brine." Now let us consider the following examples:

1. ("rine" $ fname)—this expression evaluates to TRUE as the variable fname contains the string "rine."
2. ("rine" $ gname)—this expression also evaluates to TRUE as the variable gname contains the string "rine."
3. ("rine" $ hname)—this expression evaluates to FALSE as hname does not contain the string "rine" in it.
4. ("rine" $ xyz)—this expression evaluates to TRUE as the variable xyz contains the string "rine."

As you can see, this operator locates all names that contain the specified string. In most programming languages, the shorter string is located on the left-hand side of the relational operator, but it is possible that some language could have the shorter string on the right-hand side too.

Relational expressions are used for decision making and are used in control statements. Relational expressions form part of logical expressions. But relational expressions are not used in assignment statements. Most logical errors in computer programs stem from poorly formed relational expressions.

Best Practices in Forming Relational Expressions

The following are the best practices in forming relational expressions:

1. When using numeric data in relational expressions, it is better to ensure that the arithmetic expressions on both the sides of the relational operator are of the same precision. That is, compare integer to integer, single-precision to single-precision and double-precision to double-precision.
2. It is better to use variables in relational expressions than arithmetic expressions. If it is unavoidable to use expressions, keep the expressions small so that it is easy to understand and debug programs.
3. When comparing strings of characters, it is better to convert all the characters to either uppercase or lowercase before the comparison. Of course, this does not apply to cases like checking the passwords where the case of the characters is significant.
4. It is always better to enclose the relational expression in a set of parentheses. It removes any confusion in debugging the programs.

Logical Expressions

Computers are built using integrated circuit chips (IC), which have transformed into large scale integrated chips (LSI chips) and now into very large scale integrated circuit chips (VLSI chips). These chips are built based on logic circuits referred to as gates, namely the AND gate, OR gate, and NOT gate. These gates are built using diodes and transistors. While a detailed explanation of these hardware components is out of scope for this book, it is necessary for us to understand the basics of these chips, so we have a clearer understanding of the logical expressions we use in computer programming. These gates are depicted in Figure 7.1.

These gates can have multiple inputs but have only one output. The combination of inputs and the corresponding output is depicted in a table referred to as the chip's "truth table." NOT would have only one input and one output. The truth table for the gates shown in Figure 7.1 is depicted in Table 7.2. In the field of electronics, the value TRUE is depicted as 1 and FALSE is depicted as 0.

Now the explanation is as follows. The AND gate produces an output only when all the inputs are present. In other words, the AND gate produces a TRUE output only when all the inputs are TRUE. The OR gate produces a TRUE output when any one of the inputs is TRUE. The OR gate produces a FALSE output only when all the inputs are FALSE. The NOT gate produces an output that is opposite of the input. That is, if the input is TRUE, the output would be FALSE, and if the input is FALSE, the output would be TRUE.

In logic gates, there are other gates built using these basic gates. They are NAND, NOR, and XOR gates. A NAND gate is a combination of an AND gate and a NOT gate. That is, the output of an AND gate is fed as input to a NOT gate. A NOR gate is one in which the output of an OR gate is fed as input to a NOT gate. An XOR gate (or Exclusive OR gate) is a special variety of OR gate. A representation of an XOR gate is depicted in Figure 7.2.

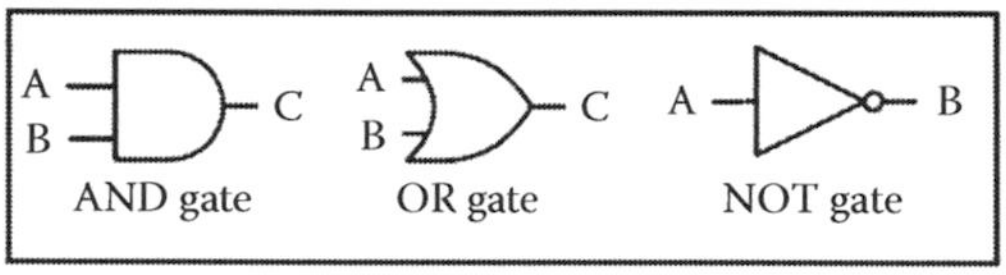

FIGURE 7.1
Logic gates.

TABLE 7.2

Truth Table for Logic Gates

Gate	A	B	C
AND	TRUE	TRUE	TRUE
	TRUE	FALSE	FALSE
	FALSE	TRUE	FALSE
	FALSE	FALSE	FALSE
OR	TRUE	TRUE	TRUE
	TRUE	FALSE	TRUE
	FALSE	TRUE	TRUE
	FALSE	FALSE	FALSE
NOT	TRUE	FALSE	—
	FALSE	TRUE	—

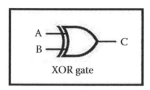

FIGURE 7.2
XOR gate.

XOR gate is like an OR gate in its input/output combinations except when all the inputs are TRUE, it returns a FALSE output. That is, an XOR gate returns a TRUE output only when any one of its inputs are TRUE but not when all inputs are TRUE.

These logic gates are the basic components of our electronic computers and also are the basis of our logical expressions.

Now returning to our discussion on computer programming, the inputs are relational expressions and the output is the result of the evaluation of the logical expression. A logical expression is a combination of expressions joined together by logical operators. Unlike in relational expressions, we can have more than one logical operator in a logical expression. The logical operators are AND (&&), OR (||), NOT (!), and XOR. XOR is a rarely used in the computer programming fraternity and not many programming languages support it.

All the logical operators have the same precedence level, so the logical expressions are always evaluated from left to right. The expressions placed on either side of the logical operator can be either relational expressions or variables of data-type Boolean. Let us assume that a, b, and c are expressions valid for use in logical expressions. Now let us consider some examples:

1. a && b—Now the output would be as follows:
   a. When both expressions evaluate to TRUE, the delivered result would be TRUE.
   b. In all other cases, the output delivered would be FALSE.
2. a || b—Now the output would be as follows:
   a. When both expressions evaluate to FALSE, the delivered result would be FALSE.
   b. In all other cases, the output delivered would be TRUE.
3. a && b || c—This expression first evaluates (a && b), and the result would be used to logically relate to c. The results would be as follows:
   a. When a, b, and c evaluate to TRUE, the result would be TRUE.
   b. When a evaluates to FALSE, but b and c evaluate to TRUE, the result would still be TRUE, as c is connected to the other part of the expression by the OR operator.
   c. When a evaluates to FALSE and b also evaluates to FALSE but c evaluates to TRUE, the output would be still TRUE because one expression of the OR operator is TRUE.
   d. When a is TRUE but b is FALSE and C is TRUE, the output would be still TRUE because one expression of the OR operator is TRUE.
   e. When all three expressions, a, b, and c, evaluate to FALSE, the delivered result would be FALSE.

4. !a—This is an example of using the NOT operator. The NOT operator returns a value that is opposite to the value delivered by the evaluation of the expression. The expression evaluates as follows:

   a. If the expression a evaluates to TRUE, the delivered result would be FALSE.

   b. If the expression a evaluates to FALSE, the delivered result would be TRUE.

We will see some examples of how all these expressions are used in our chapter on control statements.

Best Practices in Using Logical Expressions

While it is not really necessary to use parentheses in logical expressions, it is better to enclose a logical expression in a set of parentheses. Logical expressions would invariably be used in control statements and parentheses would set the logical expression as a separate entity in the statement. This helps in understanding the logic of the program, especially when debugging and during program maintenance.

A logical expression would invariably contain a minimum of two relational expressions. It is always a good practice to enclose each relational expression in a set of parentheses so that the expressions evaluated by the logical operator clearly stand out from each other. This facilitates easy understanding of the logic of the program and easy program maintenance.

While we can have any number of logical operators in a logical expression, it is better to restrict the number of logical operators in a logical expression to a maximum of three. If we use more than three logical operators in a logical expression, it would be pretty tough to decipher when debugging the program or during program maintenance.

8

Control Statements

Introduction

When a program is executed, the computer begins with the first statement and moves towards the last statement, executing all the statements sequentially one after the other. The execution of a program is akin to a waterfall. In a waterfall, the water, once it begins falling, will not stop until it touches the ground and then it would flow, taking the downward slope. Similarly, once a computer begins executing a program, it will not stop until it reaches the last statement, which informs the computer that the program has ended and instructs the computer to take all necessary actions to stop the execution in a smooth manner and release all the resources held by the program.

But often, we cannot allow the program to execute like a waterfall. We need to take some programmed decisions, and move the program execution depending on the outcome of a programmed decision. Control statements are the tools by which we can control the program flow as we desire based on the programmed decisions we built into the program. As the name implies, these statements control the flow of program execution.

We have the following statements that assist us in controlling the flow of program control:

1. Goto
2. If ... Then ... Else
3. Switch ... Case
4. Loops
   a. Counting based loops
   b. Condition based loops
      i. Condition checked at the beginning
      ii. Condition checked at the end

Let us discuss each of these statements in the coming sections.

Goto Statements

These statements are the initial control statements, and, in fact, this is the statement at the backend of all control statements. This statement is also called the branching statement. Once this statement is executed, the program flow is prevented from going to the next statement and is branched off to a different predefined statement in the program.

Goto statement can be used in a standalone manner or as part of an If statement. A Goto statement pushes the execution to a different statement and leaves it at that. The Goto statement usually consists of the keyword "Goto" followed by a label or a statement number. The label indicates the statement to which the control needs to be passed on to. Then the statement that receives control needs to be prefixed by the label. In some programming languages, each statement is sequentially numbered. In such languages, the keyword Goto is followed by the desired statement number. In other programming languages, the method of declaring and using statement labels is described in the language specification. Generally, the syntax of a Goto statement looks like:

Goto Label/Statement Number

The syntax would be the same whether it is used in combination with an If statement or in a standalone manner. As experience in programming is gathered, the disadvantages of using simple Goto statements came to the surface. Once the control is shifted by the Goto statement, the execution follows the waterfall method once again. Unless the programmer is very diligent, the programming can branch off to an undesired location. To bring back the flow of program execution, we need to use another Goto statement. But the major disadvantage arises when the program is maintained. When a program is modified, it is necessary to search out all the Goto statements and change the label/statement number at every occurrence of the Goto statement. Even while programmers are very diligent, some Goto statements are missed and it can cause problems. Not only that, it can be very difficult to locate the source of the problem and correct it.

Therefore, the use of Goto statements in programming has been restricted severely. In fact, the use of Goto statements for branching off execution is completely done away with. Now they are useful only for trapping unforeseen irrecoverable errors and closing the program smoothly. Except for error-trapping, it is not suggested to use the Goto statements in current programming practices. Every programming language provides for a Goto statement only for an error-trapping purpose. The developers of programming languages have not so far been able to come up with a better alternative to the Goto statement for handling an unexpected and irrecoverable error.

So, I too advocate that you do not use the Goto statement for any purpose other than error-trapping in your programs.

If ... Then ... Else Statements

This statement is the most useful control statement for programmed decision making. The If statement makes use of a relational or logical expression to make the decision. At a minimum one relational expression is essential for an If statement. In some programming languages, the entire If ... Then ... Else ... statement is written on the same line. But in most programming languages, the If statement spans across multiple lines. It generally takes the form:

```
If <relational or logical expression> Then
   Statement
   Statement
   ...
   ...
Else
   Statement
   Statement
   ...
   ...
Endif
```

In the programming languages of the C-family, they use curly braces ({}) to enclose the statements between Then ... Else and between Else ... and Endif. They also do not use an Endif statement at all, and they simply begin and end the If statement with a set of curly braces.

Often in real life, decisions are not simple to make with just one expression. We may need to use multiple expressions. In such cases, we may need to use multiple If statements to make the decision. For example, consider this classification of people by their age:

1. If the person is between 5 and 12 years old, that person is a child
2. If the person is between 20 and 65 years old, that person is an adult

Now, we use these rules for selecting a person's stage in life. Now we need to develop a set of control statements to make the right decision. Let us assume a variable named "age" to contain the value of the age of the person between 1 and 100. With this let us form a set of If statements to arrive at the right decision.

```
If the (age >= 5) and (age <= 65) Then
   If the (age <= 12) Then
      Specify the person as a child
   Else
      Specify the person as an adult
   Endif
Else
   Specify the person as not categorized
Endif
```

Now, as you can see, we have an If statement as part of another If statement. This type of embedding an If statement within another If statements is referred to as "nesting" of the If statements. In the present case, there are just two levels of If statements. We refer to this as "the nesting of If statements is two levels deep." Real life provides us instances wherein the nesting of If statements go even deeper than two, but if we nest the If statements too deep, it becomes very difficult to analyze the programs to debug errors. So, the general rule we follow in the programming fraternity is to limit the nesting of If statements to a maximum of 3.

Here are the general rules for writing If statements:

1. An If statement must have at least one relational expression in it.
2. An If statement can have multiple relational expressions joined by logical operators.
3. While there are no restrictions on the length of an If statement, it is recommended to restrict the length of an If statement to the length of a line permitted on the computer screen so that it can be read easily without scrolling the screen horizontally. If an expression needs to be longer than the length of a line, it is better to use assignment statements to a Boolean data-type variable in a preceding statement and then use it in the program.
4. While there is no restriction on the number of expressions in an If statement, it is suggested to restrict the number of logical expressions to 2. That is, a total of four relational expressions as each logical expression would have two relational expressions. This facilitates easy analysis of the statement while debugging the program. In case the decision needs more than two logical expressions, it is better to break the expression into more If statements and nest them or use a Boolean data-type variable to assign the results of the evaluation of the expression and then use it in the If statement.
5. Most programming languages allow writing the If statement in multiple lines and it is recommended to write an If statement in multiple lines, even if it can be accommodated in a single line. This enhances the clarity of the program and makes it easy to analyze and understand the program during its debugging and maintenance.
6. It is possible to ignore the Else part of the If statement in some cases, but real life is so weird that the data can throw up surprises and the Else statement may be needed. So, it is advocated to include the Else part of the If statement in every case, even if no statements are embedded between the Else and the Endif keywords.

The If statement is perhaps the most important of the computer programming skills and mastering it is essential to writing good quality computer programs.

Switch ... Case Statements

A Switch ... Case statement is a control statement in which control is switched to a practically unlimited number of branches based on the value of a single variable. Each value of the switching variable would be branched off to a specified set of statements. When the value of the variable does not match any of the defined cases, the control is switched off to a set of statements defined under a special case, "default." Default is part of the Switch ... Case statement. The syntax of the Switch ... Case statement is as follows:

```
Switch <Variable>
    Case a
        Statement
        Statement
    Case b
        Statement
        Statement
    Case n
        Statement
        Statement
    Default
        Statement
        Statement
Endcase
```

In the earlier statement, a, b, and n are the values for the variable specified in the Switch statement.

Here are the rules for writing the Switch ... Case statement:

1. The Switch case consists of a switching variable defined by the Switch keyword.

2. There would be a number of cases based on the value of the variable. Each value is defined by the Case keyword.

3. The Case keyword is usually followed by a constant, which can either be numeric or character data.

4. Some programming languages allow a range of values with a special syntax.

5. Generally, expressions are not allowed in the Case statements, but exceptions can always arise.

6. Each Case statement includes a number of statements. There is no restriction on the number of statements that can be under a Case statement.

7. The last case for this statement would be the Default statement, which specifies the action to be taken when the value of the variable does not match any value specified in all the Case statements.

8. The statements in the Case and Default statements can be action statements or branching statements.

9. Most programming languages require a statement to denote the end of the Switch ... Case statements. Some would define a keyword, such as Endcase. Some programming languages, such as in the C-family, just use a closing curly brace "}" to denote the ending of the Switch ... Case statement.

10. The last of the statements under each Case would usually be a "Break" or "Exit" statement. It would be usually the only word in that statement. The execution of Break (or Exit) statement would take the execution to the statement that immediately follows the Endcase statement.

The total set of statements, beginning with the first Case statement and ending with the Endcase (or its equivalent) statement, are usually called the Switch Block. The statements in each of the Case statements are usually called the Case Block.

Usually the conversion of numbers to strings of characters or vice versa is achieved by a Switch ... Case statement. Let us consider an example for converting month number to month name:

```
Switch MonthNumber
    Case 1
        MonthString = "January"
        Break
    Case 2
        MonthString = "February"
        Break
    Case 3
        MonthString = "March"
        Break
    Case 4
        MonthString = "April"
        Break
    Case 5
        MonthString = "May"
        Break
    Case 6
        MonthString = "June"
        Break
    Case 7
        MonthString = "July"
        Break
    Case 8
        MonthString = "August"
        Break
    Case 9
        MonthString = "September"
        Break
    Case 10
        MonthString = "October"
        Break
    Case 11
        MonthString = "November"
        Break
    Case 12
        MonthString = "December"
        Break
    Default
        MonthString = "Unclassified"
        Break
Endcase
```

The Break statement in the Default block is really superfluous, as the program execution falls to the next statement, which in this case is the Endcase statement, but it is a good practice to write the Break statement. Sometimes during program maintenance, we are likely to add another Case statement after the last Case block. Software packages such as the word processing packages make heavy use of the Switch ... Case statements. Some programmers consider the Default statement as unnecessary and avoid it, but it is a good practice to include it. Real life throws unexpected data that can never be completely guessed at the time of writing the program, so the Default statement handles all exceptions that real life throws at the program.

Utility of Switch ... Case Statement

Switch ... Case statement is used when there is a need for making a decision based on one condition that can have multiple outcomes. We use If statement when the decision has one or two possible outcomes. When the possible outcomes are numerous, we use Switch ... Case statements. When converting from numbers to strings, this statement comes in handy. Switch ... Case statements also come in handy when we are developing software packages such as MS-Word, where in the character being typed in can lead to any one of a variety of different actions.

Precautions for Using Switch ... Case Statement

It is essential to use the Break (or Exit) statement at the end of the statements in every Case block. Most often, programmers tend to neglect the Break statement in the last block of Case statements. It is also common to forget the Break statement in the Default block of statements. Break forces the execution to go outside the Switch block of statements. While writing computer programs, we need to remember that the program will certainly undergo modification or enhancement. We are likely to add a Case block after the last Case block, in which case the absence of the Break statement can cause havoc.

Another mistake programmers often commit is forgetting to include the Default block. They possibly think that all possible outcomes of the decision scenario are covered, but real life is so weird that it can throw up incredible values at the programs resulting in unexpected aborts. So, it is essential to include the Default block in every Switch ... Case statement.

Loops

A Loop is a construct that causes a set of statements to execute iteratively. How many times do the statements in the loop block execute? It depends on the type of loop that we constructed. Of course, the loop has to be finite and if the loop becomes infinite, then we have inserted a bug and the program freezes and needs to be manually interrupted. Loops are basically two types:

1. Counting-based loops
2. Condition-based loops
   a. Condition checked at the beginning
   b. Condition checked at the end

Counting-Based Loops

Counting-based loops execute the set of statements for a fixed number of times based on the definition at the beginning. Most modern programming languages use the For ... Next construct for using this type of loop. "FOR" loops are the primary tool in handling arrays in programs. Generally, the FOR loop has the following syntax:

```
For i = j to k Step 1
    Statement 1
    Statement 2
    Statement n
Next
```

In the earlier example, i, j, k, l, and n are numeric variables. The variable j denotes the initial number at which the loop begins, and k is the last number at which the loop stops. Once the value of i becomes greater than k, the execution of the loop stops. The keyword "For" signifies the beginning of the loop. The keyword "Step" indicates the value by which the variable i would be incremented after every iteration. The keyword "Next" indicates incrementing the value of the variable i by the amount of the value indicated after the keyword Step.

The variable i would assume an initial value of j and executes the statements in the loop block. The set of statements in the loop block would be executed even when the value of the variable i becomes equal to k. Therefore, the number of times the set of statements, in the loop block, are executed, is equal to (k − j + 1). For example, if the statement is "For i − 1 to 6 Step 1," then it would be executed (6 − 1 + 1 = 6) times. Let us consider an example. Let us assume a single-dimensional array having six cells. Let us fill this array using a FOR loop:

```
For i = 0 to 5 Step 1
    Read m
    arr(i) = m
Next
```

In the earlier example, i is a counting variable and m is the input variable. Note that we began the loop with the initial value of zero. Generally, the first element of an array is denoted as 0th element/cell. With the value of i being incremented by 1 in each of its iterations, it assumes the values of 0, 1, 2, 3, 4, and 5. As you can see, the loop would be executed six times to fill the array with values received as input. In each of the iterations, the computer performs two operations; namely, it receives one value as input and fills the corresponding array element with that value. When i is incremented and reaches a value of 6, the execution of the loop stops and moves to the next statement in the program.

Let us now consider a two-dimensional array and fill it with values input by the user. Let arr be a two-dimensional array with 3 columns and 6 rows. Here is the pseudo code for it:

```
For i = 0 to 5
    For j = 0 to 2
        Read m
        arr(i, j) = m
    Next
Next
```

In the earlier example, i and j are counters. The variable m is the input variable. The first FOR loop sets the value of the row of the array into which the values are being filled. The second FOR loop sets the value of the array column for filling the value. The statement arr(i, j) indicates the array element where i is the row number and j is the column number. If i = 2 and j = 2, then the value goes into row #3 and column #3 (as the computer counts from zero and we count from 1!).

Here are the general rules for the FOR loop.

1. FOR loops are used to define loops that are executed a finite number of times based on the value of the counting variable.

2. A FOR loop needs a counter, a numeric variable used for counting the number of iterations executed by the loop.

3. It also needs an incrementing value/variable to increment the counter by after completing every iteration. If we do not define an incrementing variable, most computers assume the increment to be 1.

4. The FOR loop needs some indicator to indicate the last statement in the loop. Some programming languages use Next, and some simply use a closing curly brace—"}."

5. The FOR loop can be blank, that is, there need not be any statements between the For statement and the Next statement. Such loops are called as delay loops and are used sometimes in the programs. It just delays the program execution for the time it spends in counting the numbers specified in the loop.

Utility of FOR Loops

FOR loops are used mostly in programs solving mathematical problems. Matrix algebra, factorials, prime number generation, random number generation, and others that require the repeated execution of a set of statements when the number of times those statements need execution is known beforehand. Of course, these can be used in other areas too, but the prerequisite to use this construct is that we need to know the number of times the set of statements needs to be executed.

Precautions for Using the FOR Loop

While it is common to use numeric constants in the FOR loop for the counter and the step, it is better to use variables and input the values using a data file. This helps in software maintenance when the values need change. Then we can simply change the data file without resorting to a program change, which is more tedious than changing the data.

Condition-Based Loops

In these loops, the set of statements in the loop block are executed as many times as the loop condition is either TRUE or FALSE as programmed. This loop is used when we do not know or cannot know how many times the set of statements needs to be executed by the computer. One example that readily comes to mind is processing the records from a data file or a table. We normally would not know the number of records contained in the file or table. Then we define a loop using a condition. These loops are basically of two types:

1. *Condition checked at the beginning*: In this loop, the condition is checked before entering the loop and executing the statements contained within the block. If the condition evaluates to FALSE (or TRUE as the case may be), the statements in the block do not get executed even once.

2. *Condition checked at the end*: In this loop, the condition is checked in the last statement of the loop statement. Therefore, the all the statements in the loop block get executed at least once in this loop.

WHILE Loop

The While statement is the most versatile construct for the loop that checks the condition at the top of the loop. Do ... While is the popular construct for the loop that checks the condition at the bottom of the loop block. Of course, we need to check the programming manual of the specific programming language we are using when writing the program to know the exact constructs for these loops. Let us see some examples of the While loop. Let us assume FileofNames as a data file or a table from which we read records and simply print the name.

```
While EOF != TRUE
    Read a record from FileofNames
    Print the name
Wend
```

EOF, meaning "end of the file," is generally a variable to indicate if all the records in the file or table are processed and that there are no more records in the file or table to be processed. This is the general notation used by many programming languages to denote the condition that there is no more data available in the file or table. From the earlier example, we note the following aspects:

1. As long as there are records in the file or table, the condition evaluates to TRUE. It goes something like this:
   a. When there are records, the answer to the question "is EOF NOT TRUE" would be "Yes, EOF is NOT TRUE." The execution falls to the first statement in the loop block.
   b. When all the records are exhausted, the answer to the "is EOF NOT TRUE" would be "No, EOF is TRUE." The execution falls to the first statement coming after the loop block.

2. When all the records in the file or table are exhausted, the condition in the While statement evaluates to FALSE and the computer exits the loop and begins execution of the statement that immediately follows the loop. We can also insert an IF statement in the loop and check further conditions, along with a BREAK or EXIT statement to exit the loop. Let us consider another example.

```
While EOF != TRUE
   Read a record from FileofNames
   If (name = "John Doe" Then
      Print the name, phone_number, address
      EXIT
   Endif
Wend
```

In this example, we added a condition to check the name is John Doe, and if it is, then exit the program. That is, we are trying to locate the contact details of a specific person from the file or table.

The WHILE loop is a very versatile one, and it can replace the FOR loop. Let us rewrite the segment of code that we wrote earlier for filling a single-dimensional loop in the section on FOR loops. Look at the following example:

```
i = 0
WHILE i < 6
   Read m
   arr(i) = m
   i = i + 1
WEND
```

It will do the same operation as the FOR loop example given in the earlier section on FOR loops. We had to add two more statements: one to initialize i and one to increment i. But, one question: why should we use a WHILE loop where a FOR loop is the perfect fit? The answer is, we need not. I just showed the possibility. I have seen some programmers prefer to only use the WHILE loop to the exclusion of all other loops. Here are the general rules for writing WHILE loops:

1. A WHILE loop consists of a minimum of two statements, one to initiate the loop and the other to end the loop. The keyword WHILE is the construct preferred by most programming languages; the keyword WEND is not very popular. In the C-family of programming languages, a closing curly brace "}" is used for ending the loop.
2. The WHILE statement needs to have a relational or logical expression along with it. This condition sets the rule for stopping the execution of statements within the block and passes on execution to the statements following the loop block.
3. There can be any number of statements embedded between the WHILE statement and the WEND statement.
4. There must some statements contained within the loop block that cause the condition to evaluate to FALSE. Otherwise, the loop becomes infinite and freezes the program and the computer.
5. Generally, the WHILE loop is not used to branch control to other statements in the program. The IF statement is the appropriate one for that purpose. But unexpected errors can crop up that we need to trap using appropriate statements. Only in such cases would we use EXIT or BREAK to exit the WHILE loop.

Precautions for Using the WHILE Loop

The "WHILE" loop is a very useful construct in computer programming. It is used frequently in programs. The mistakes often committed by programmers are mainly two.

One is to forget to insert statements that terminate the loop by making the condition evaluate to FALSE. This causes the loop to execute infinitely. The second mistake is to forget to catch unexpected errors. If we are careful about these two aspects, we can write impeccable WHILE loops in our programs.

DO ... WHILE Loop

This loop is almost identical with the WHILE loop except that the condition is checked at the bottom of the block. The syntax of this statement is as follows:

```
DO
    Statement 1
    Statement 2
    Statement n
WHILE <relational or logical expression>
```

As you can see, the WHILE statement is at the bottom of the block. The DO statement does not require any other supporting keywords or expressions. It indicates the beginning of the loop. The WHILE statement needs the condition defined by a relational or logical expression. All other aspects of this loop are same as the ones detailed for the WHILE loop in the preceding section. Therefore, I am not repeating them here once again.

Other keywords are used in different programming languages. Notable among those keywords is the REPEAT ... UNTIL set used by the PASCAL programming language. It is also used in some other languages. As there are now a plethora of programming languages, perhaps there could be different keywords for this kind of loop. It is also possible that a few differences could be there between what I described earlier and what those languages prescribe.

Best Practices in Programming Loops

Loops are a very important component of computer programming. I would go on a limb and say that it is impossible to write any meaningful program in real life without using loops, especially when you read data from a file or table, as well as when you write data into a file or table. Therefore, a serious programmer ought to master the programming of loops and write flawless programs. Here are some best practices while writing loops in programs:

1. First and foremost, we need to ensure that the loop would never be an infinite loop. I am sad to say that this is one of the biggest challenges in the debugging and maintenance of computer programs. Computer freezes occur due to infinite loops. We must build in condition statements in such a way that it becomes TRUE and terminates the loop. Ensuring that the loop terminates is a best practice in programming loops.

2. It is often better to use FOR loop as much as possible, as it is more efficient than the WHILE loop.

3. While programming the FOR loop, I suggest that the maximum limit in the statement be read from a file rather than hard-coding it (writing a number inside the program), as it would avoid program change during software maintenance. We need to accept the reality that in this world, everything that can change, will. Changing a data file is much easier than changing the code, compiling it, testing it, debugging it, linking it to libraries, and moving it to the production environment.

4. Most programming languages provide for composite keywords that shorten statements. These keywords, while reducing the chore of typing, are hard to decipher during software maintenance. Therefore, I advocate writing multiple simple statements, even if it increases the typing load. It would help greatly during the software maintenance.

I avoided discussing the IF ... GOTO loops as it has disadvantages, especially in leading to error conditions. In the present day, most quality, conscious organizations prohibit using the GOTO statement except for error-handling situations. So, my humble suggestion is that you avoid the temptation of using IF ... GOTO loops in your programs.

9

Input Statements

Introduction

As we noted in Chapter 1, computers are data processing tools. Programs help the computer to process the data, and input statements are the statements that bring data from the outside world into the computer. In this chapter, we look at the way by which computers take in data from the outside for processing it and delivering the output.

How Data Comes into the Computer

How does a computer receive input data from the outside world? Simple: a computer receives input using an input device. Here are the steps in the computer receiving data:

1. The computer receives data from the outside world only when a program executes an input instruction.
2. The input instruction tells the computer:
   a. The specific input device from which data is to be received.
   b. The data that needs to be brought in from the specified device.
   c. The locations in RAM where the received data is to be stored.
3. The computer receives the data and stores it at the specified location.

This is the manner in which data from the outside world is input into the computer. The following actions are performed by the computer in receiving data from the specified input device:

1. The computer loads the device driver of the specific device from which data is to be received into the RAM. A device driver is the program that performs the following actions:
   a. It controls the device and issues commands to it in a manner that the device understands the instructions and performs the required actions.
   b. It checks if the device is connected, powered up, and ready for interaction with the computer.
   c. It checks the device for proper functioning and raises error messages if it is not functioning.

    d. It translates the commands from the computer into equivalent commands of the device and vice versa. The commands of the computer and the input device are in their respective native languages. The device driver acts as an interface between the two, translating the commands from both sides so that a continuous communication is established between the computer and the input device.

    e. It establishes communication between both the device and the computer.

    f. It receives data from the input device, translates it to suit the computer, and delivers it to the computer.

2. The operating system executes the device driver and gets the data into the RAM.

3. While the data is being received, the operating system deallocates the CPU from the program and allocates it to another program. The program is set to the "wait" state and is moved to the RAM.

4. Once the required data is received, the operating system changes the state of the waiting program to the "ready" state and allocates the CPU to it, adhering to the algorithm for CPU scheduling and allocation.

5. The program begins re-execution of the waiting program once again when the CPU is reallocated to it.

We have to note here that there is a huge difference between the speed of the CPU and of any input device. It would be a travesty if we make the CPU wait all the time that the input device takes to supply the information. Therefore, we release the CPU to other tasks while the input device is taking its time to input the information. This is handled in the following ways:

1. *Use of interrupts*:

    a. The CPU places an interrupt on the input device and returns to the next task waiting for its allocation.

    b. The input device makes the input ready and places an interrupt on the CPU.

    c. The CPU responds to the interrupt, receives the information from the input device, and stores it in the appropriate memory location.

2. *Buffers*:

    a. All the I/O devices are equipped with a small amount of RAM generally referred to as "buffer," as it is used to provide a buffer to bridge the gap in the speed difference.

    b. When the CPU places an interrupt on the device, the device fills this buffer and places an interrupt on the CPU, which reads the input from the buffer of the device.

3. *Direct memory access (DMA)*: Usually, the receiving of data from input devices is not handled by the main CPU. Most computers have an auxiliary CPU, referred to as DMA (Direct Memory Access), which handles all the interfacing between the peripheral devices and the RAM. The CPU passes the I/O activity to be performed along with the ID of the device, the addresses of the RAM, and the details of the data to be received to DMA, and it performs the action of receiving the data and storing it at the appropriate locations in the RAM. Once the action is completed, it informs the operating system about the completion of the action and the CPU changes the state of the waiting program to "ready."

Most of the present-day computers use the DMA method to resolve the speed differences between the CPU and the I/O devices.

However, as a programmer, we need not be concerned with all this. All these activities are carried out by the operating system of the computer silently and in the background without requiring any intervention from the programmer or the user. Of course, the user still has the role of providing the right data to the input device.

We had enumerated various input devices in Chapter 1; now let us have a recap of a few important devices. These are:

1. *The computer keyboard*: Usually, the keyboard is the default input device to most present-day computers. Some time ago, the card reader was the default input device. In real-time systems, the machine could be the default input device.

2. *The hard disk, which is mounted internally on the computer itself*: The input from the hard disk is supplied by the data files and database tables. We really do not access the hard disk directly, but access the files and database tables on it. We receive data from data files and tables of the databases.

3. There are other devices that provide data from files including the magnetic tape, the CD (Compact Disk), and the DVD (Digital Video Disk). The devices such as the punched card reader and floppy disk have become obsolete and are not being used any more.

4. There are devices, like the machines and hardware of vehicles like cars, airplanes, and rockets that provide data on a continuous basis.

Now let us consider statements that help us in receiving input data.

Opening of Files and Tables for Input

We have noted in the preceding section the main input devices, and the input statements differ for each of those devices. Now let us look at the input statements from the default input device, namely, the keyboard. The input statements contain the following components:

1. *The ID of the input device*: This would usually be a number, at least, inside the OS. Our programming language would provide a number, either predefined or to be defined by us. This will be used by the computer to locate the device from which to obtain the input.

2. *The location of the information*: This could be a file or database table. The file and database table would be assigned numbers by us.

3. *The details of the information to be received*: These would be the variables that were defined earlier in the program by us. Variables would be identified by the names we assigned during the declaration of the variables. Along with the variable names, we also need to specify the order in which the variables are to be received. This is achieved by specifying the variables in the order in which they were stored in the file or table. If the order specified in the program is different from the order in which the variables were stored, then error would result.

We need to open the file before attempting to read from it. The following functions are performed in the opening of a file:

1. The OS would locate the device on which the file is located. This would involve checking the health of the device and ensuring that it is operational and has the specified file.

2. Copy the file information from the VTOC (Volume Table of Contents). It is also referred to as File System, File Allocation Table, or by any other name into RAM.

3. Allocate RAM to hold one record or a block/set of records in the RAM to hold the information read from the file.

4. The file structure defined in the program would be compared with the structure of the file on the disk and if they are differing, then an error would be generated by the OS that needs to be handled by the programmer. We have a separate chapter on error-handling.

5. If a filter (a condition to selectively retrieve the records) is set, apply the filter while retrieving the records. This involves allocating RAM and store the filter information in it for reference during every read operation.

6. Place an interrupt on the CPU to indicate that the input is ready.

For reading information from a database table, we need to connect to the database and then open the table. It involves the following operations:

1. Connect to the database. We need to provide the following information while connecting to a database:

   a. The opening string of the machine on which the database is located. This would be a string of characters and includes the IP address, the name of the machine, the disk volume, the name of the DBMS software (like Oracle, SQL Server Progress, and so on) on which the database is located, and so on. The contents of the database opening string depend on the installation of the hardware and software combination.

   b. The user ID and password for the database. Please note that this combination is not for the machine but only for the database that is proposed to be opened.

   c. The type of security, which can be the database administrator, user, or any specific database role that needs to be specified along with the opening string.

   d. Any lock (read lock, write lock, exclusive use, or any other lock) that needs to be applied to the database while performing database operations.

2. Once the string is provided to the connect statement, the OS will connect the program to the database.

3. The detailed information about the database, that is, its location, ID, table information, permissions, user ID and password, etc., are copied to the RAM for reference during each database operation.

With this, the database gets connected to the program and we can perform read operations on the database.

Once the database is connected to the program, we can access any table from the database and use it for read operations. Opening any table involves the following operations:

1. The table information including the ID, number of records contained in the table, the indexes generated on the table, and the primary and secondary keys would be copied into the RAM for reference during each table operations.

2. Copying the security information and security permissions into the RAM for reference during the execution of the program.

3. Copying the filter information for the retrieval of the records.

Now, the table is ready for receiving information input to the program under execution.

Input Statements

Now, most of the user input is taking place using the GUI (Graphical User Interface). In the GUI screens, we have various controls, and the important ones among these are:

1. Form
2. Frame
3. Text boxes
4. Combo boxes
5. List boxes
6. List views
7. Grids
8. Radio buttons
9. Command buttons
10. Labels
11. Links

Let us discuss each of them hereunder.

Form

In a GUI environment, all the controls are placed on the form. When the program is executed, the form is displayed on the screen along with all controls placed on it. A form has many properties associated with it, and here are some important ones:

1. *Name*: Every form has a name associated with it. We use this name in the programs when we refer to it. We use "form.control.property" to refer to the property of a control on a specific form. For example "loginform.userid.text" refers to the text entered into the box named "userid" on the form named "loginform."

2. *Icon*: This defines the picture of icon type to be displayed on the top left-hand corner of the form (or at a place defined by the programmer or the specific programming language).

3. *Caption or title*: This is the text displayed on the top border of the form when it is displayed or at an appropriate place defined by the specific programming language.
4. We can set the colors for the background and foreground.
5. *Enabled or Disabled*: By setting this property as desired, we can allow modification of the contents of the controls on the form. When we set this property to Disabled, the controls are displayed on the form but disallow entry of fresh values or modification of existing values.
6. *Visible*: By setting this property dynamically, we can make this form appear and disappear during execution as desired.
7. There are many more properties to achieve various requirements of the form.

A form has a number of events associated with it, and the important ones are enumerated here. Each of these events can be programmed as needed by us. Please note that a specific programming language may have different names for these controls:

1. *FormLoad*: In this event, we can program all those actions that need to be executed before allowing the user to use the form. All these actions would be executed by the computer as it loads the form. We usually program such actions as authenticating the user (see if the user accessing the form was properly logged in), ensuring that the user has necessary permissions to use the facilities provided for in the form, block out those specific controls for which the logged-in user has no security clearance, open a connection to the database, and so on.
2. *FormUnload*: In this event, we program all those actions that needed to be executed before closing the form and taking it off the screen. Typically, we write code to close all database connections, restoring the form that was on top before this form was loaded, closing the application if necessary, logging off the user, and so on.
3. *Activate and deactivate*: In these events, we either activate the form and bring it on top or deactivate the form and minimize it. By activation, we bring the form on top of the screen and enable it to receive inputs or use the facilities provided on the form. By deactivation, we minimize the form and disable it from receiving inputs or using any facilities provided on the form.
4. *Resize*: In this event, we change the size of the form to the desired size. Usually the form will have three sizes, namely, normal, full, and minimized. The normal size form, typically, would not occupy the entire screen. The exact size of a normal form depends on the definition provided in the programming language. The full size however, would occupy the entire screen. The minimized form would be shown as an icon on the task bar at the bottom/top of the screen and would be taken off the screen. Of course, we can also define the form size to suit our application by giving the coordinates for locating the form and the size of the form on the screen.

Frame

A frame can be viewed as a subform or a form within a form. When we use radio buttons (to be discussed in the following sections), we need to use a frame to define all the buttons as one set. A set of radio buttons within a frame are shown in Figure 9.5. We can also use frames to arrange the controls on the screen into logical groups. The advantage of using a frame is we can disable all the controls on the frame with one programming statement. We can also enable all the controls enclosed within the frame with one programming statement. The other advantage of using frames is that we can logically divide the controls on the form into logical groups to make the use of the form easier for the user. The events associated with a frame are got-focus, lost-focus, enable, disable, visible, invisible, and so on. These properties are discussed in the following section dealing with text boxes.

Text Box

A text box (Figure 9.1) is usually a rectangular area on the screen placed at a designated place to receive a set of characters as guided by a prompt beside the box. These characters can be numbers, alphabets, and other characters as desired by the user. A text box allows for the entry of values without checking the data type being entered, and it behooves on the programmer to do the type-checking. If we do not check the data type being entered into a text box, the error would be thrown up when we assign the value to a variable if there is a mismatch of the data-type. The value entered in a text box is transient. It disappears when the screen is unloaded. We can use the value entered into a text box directly in processing without assigning the value to a variable. A text box can have a name defined by the programmer. We can also use the default name assigned by the programming language. A text box has a number of properties associated with it. The important ones are:

1. *Name*: The text box has a name associated with it. We use this name when we refer to it in the program and to capture the value entered into it for processing or storing.

2. *Text*: We suffix this property to the name of the text box. This would contain the value entered by the user. We usually use this property as "text1.text," which means the value of the text entered into the text box named "text1." We can now assign this value to a variable in this manner—"NameofStudent = Text1.Text." This statement would assign the value entered into the text box, named Text1, to the variable NameofStudent. Of course, if the type of data is not same in the text box as well as the variable, an error would be thrown up.

3. *Tab Index, Tab Stop*: The Tab Index property sets the order in which the cursor comes into this box when we press the tab button on the keyboard. By using the Tab Stop property, we can make the cursor either skip this box or stop in this box.

4. *Visible*: By setting this property, we can make the text box either visible or hidden when the form is displayed.

5. We have a few properties to select the font to display the text entered by the user, the background color of the box, foreground color of the box, the size of the box, and so on. We use these properties to enhance the appearance of the text box.

6. We have a few properties to associate the box to a field in a database table so that we can import data from the database table into the box.

FIGURE 9.1
A text box.

7. *Enabled and Disabled*: We can use these properties to permit entry of text into the box or modification of the existing text in the box. When we set the property to Disabled, the text would be visible but does not allow its modification.

8. *Tool tip*: This is a facility to provide some helpful text to the user. The text we define in the tool tip control becomes visible when the mouse hovers over the box.

The text box has a number of events associated with it. We can write a program to perform desired actions when the event is triggered. Here are the important events associated with a text box:

1. *Got-Focus*: This event contains statements to be executed when the cursor position shifts into this box either as a result of pressing the tab button or clicking the left mouse button when the cursor is positioned in the box. We use this event to highlight the text already inside the box, but there are many uses for this event.

2. *Lost-Focus*: This event contains statements to be executed when the cursor shifts out of this box. Usually, we use this event to validate the value entered into the box., but we can also program other statements as required by the situation at hand.

3. *Click*: We can include such statements that need to be executed when the left mouse button is clicked when the cursor is placed in the text box.

4. *Double-Click*: We can include such statements that need to be executed when the left mouse button is double-clicked (the left button of the mouse is clicked twice in quick succession) when the cursor is placed in the text box.

5. *Change*: This event is triggered when the value in the text box is changed. A change can happen only when there is some text already inside the box before the focus is shifted into this box and then it is changed. The Keypress event is triggered whenever a key on the keyboard is pressed, but the Change event is triggered only when the focus is shifted out of this box. We can program all such statements that enumerate the actions to be performed by the computer when this event is triggered.

6. *Keyup, Keydown, Keypress*: Keydown is triggered when the key is pressed but not released. Keyup is triggered when the pressed key is released. Keypress is triggered when the key is pressed and immediately released. We use the KeyPress event to check if the character is either numeric or non-numeric, especially when we expect the data to be numeric. We also use this event to ensure that the character is a permitted character and prevent special characters to be entered along with the Alt and Ctrl keys.

7. *Mouseup, Mousedown, Mousemove*: These events are used to program the movements of the mouse and the click of its left button. Oftentimes, we may click on a box by mistake. In such cases, we can use the Mousedown event to program what can and cannot be done. Similarly, we can use the Mouseup event to program when the pressed left button of the mouse is released. Mousemove event is used to program what needs to be accomplished when the cursor is moved by moving the mouse such that the cursor hovers on the box. Usually we highlight the text in the box when the cursor hovers over the box by programming the event.

8. *Dragdrop, Dragover*: These events are used to program the drag and drop operation during the cut and paste actions.

9. There are some more events associated with a text box, and these events are continuously enhanced as well as supplemented to make the programs more powerful, aesthetically more appealing, and reduce the actions to be taken by programmer. So, we need to refer to the manual of the specific programming language we are using for complete set of events and their functionality.

All these events can be set using the facilities provided by the IDE (Interactive Development Environment) of using program statements for the event. The other controls have similar properties and events associated with them. As I discussed these events here in detail, I would not repeat them when discussing other controls to avoid duplicating the explanation.

Combo Box

A combo box is in fact a combination box combining a text box with a list box (Figure 9.2). It is a rectangle with a downward-pointing triangle indicating selection. It is used for selecting a single value from a set of values already available in the combo box. When the program is under execution, clicking the inverted triangle results in the box being extended downwards displaying the values available for selection as shown in Figure 9.3.

A combo box can be filled with values of a specific attribute (variable or field), and then the user can select any of the values in the list. Alternatively, the user would be allowed to enter a fresh value into the box. The box provides the facility to block the user from entering a new value. Often, this would be used to take in the primary key from a database so that if an existing value is entered, the relevant data would be retrieved and displayed in the other appropriate controls. Of course, all these actions need to be programmed by the programmer! We can store the data in a combo box either in a sorted order or in their original order. Combo boxes allow the user to select the values by clicking the item in the displayed list or by pressing the first character in the item and then using arrow keys. We use combo boxes extensively to present a choice of items and allow the user to select a single existing value or to enter an altogether new value. A combo box is both an input as well as an output. It is used as output when we fill the combo box with values, and when the user selects an item, it will be used as an input.

FIGURE 9.2
A combo box.

FIGURE 9.3
A combo box after clicking the inverted triangle.

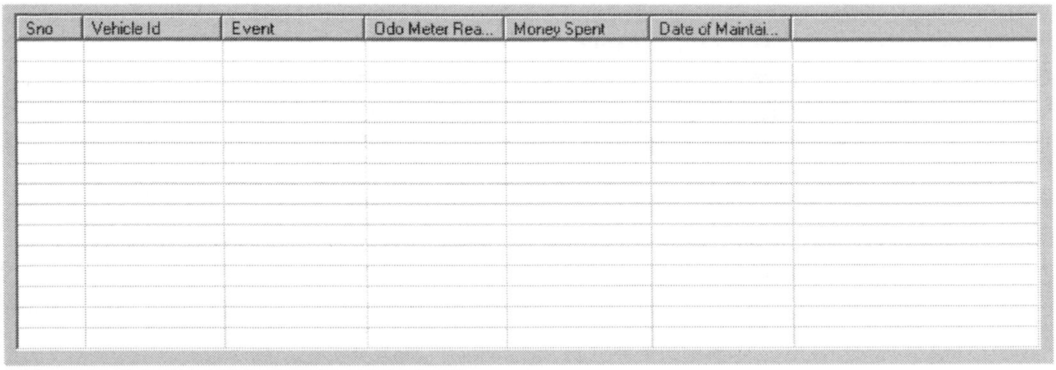

FIGURE 9.4
Grid.

List Box

While the text box and the combo box are used to receive input of one data item, the list box can receive multiple data items for input. But in both the cases, all three, text box, combo box, and the list box, receive inputs for the same attribute (or field of a database table). The list box looks exactly same as the text box with sliding bars at the right side. It permits the selection of multiple items from the displayed list. The selection of multiple items is facilitated by holding the control (CTRL) key and then clicking on the desired items.

Grid

While text box, combo box, and list box receive values for one single attribute (or a database field), a grid can receive data items for multiple attributes (or database fields or an entire record). List view, or a grid (Figure 9.4), is a table with rows and columns displayed on the screen. It can display values of multiple attributes (or fields of a database table). It can be directly connected with a database table, and each column can hold the values of a field. Each row can hold the values of a record. We can fill the grid with only the required data. In some grids, the act of changing the values in the grid would automatically change the values in the database table. A grid can be used both as an input and an output. When we fill the grid with the data from a table, it is an output, and when we receive new values from the grid, it is used as an input. We use grids in input screens in which we receive multiple data items to show that the entered data is indeed saved by displaying them in the grid. It is also used in the enquiry screen to show multiple items that resulted from the query passed on to the database table.

The grids allow for column headings to be placed on the top row. Some grids allow for sorting the entire grid by the values of a selected column by clicking on the header row at the desired column. Grids also allow us to search within the grid to locate the desired values.

Radio Buttons

Radio buttons are shown in Figure 9.5. Radio buttons are to be embedded inside a frame as shown in Figure 9.5. Radio buttons are used to present the user with a set of mutually exclusive options for an attribute from which the user can select only one option. Each radio button presents one option and together, a set of radio buttons presents all available options

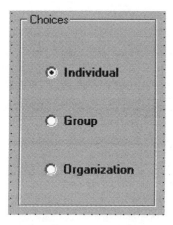

FIGURE 9.5
Radio buttons.

from which only one can be selected. All the options are visible on the screen. The same functionality can be achieved with a combo box also, but the combo box presents only one option at a time. To see the other options, we need to click the inverted arrowhead on the combo box. With radio buttons, all the available options are present on the screen itself, enabling the user to view all the options at one time without clicking anything. When the program is executed, only one radio button in the set can be selected and all others would be deselected. That is, the radio buttons are mutually exclusive. When a radio button is selected, the previous selection would be removed. Radio buttons are to be used to ensure that only one option is selected. Each radio button results in a Boolean input that is 0 or 1, or Yes or No.

The caption property is available to controls like the radio button, check box, command button, the label, etc. This is displayed along with the control when the program is executed. This helps the user in efficiently entering the needed input with ease.

Check Boxes

Check boxes are similar to radio buttons in the sense they present all the options available just like the radio buttons, but the comparison stops there. Each check box results in a Boolean input as 0 or 1, or Yes/No, for one data item. All the radio buttons in a frame are intended to obtain input for one data item whereas each check box obtains input for one data item. Check boxes are mutually inclusive, that is, the user can select multiple check boxes, all of which are valid for the scenario. Radio buttons are used to enable the user to select one option out of a set of mutually exclusive options, and check boxes enable all the options that complement and supplement each other. A set of check boxes are shown in Figure 9.6. Another aspect is that the check boxes need not be embedded inside a frame, as they are mutually inclusive.

Check all that applies

☐ School   ☐ College   ☐ Post Graduation   ☐ Doctorate

FIGURE 9.6
Check boxes.

FIGURE 9.7
Command buttons.

Command Button

Command buttons are used to indicate that the inputs are given, and that the computer can begin to process the inputs. A sample command button is shown in Figure 9.7. Each command button will usually require a program to process the data entered into the controls by the user. It initiates the execution of the program. Usually, command buttons are used to save the data, cancel the action, delete a set of data, or any other affirmative actions. It generally has the captions of Save, Cancel, OK, Yes, Book Ticket, Cancel Ticket, Transmit, and so on. Programs can make the command buttons visible or invisible to implement security based on the permissions the user has. The properties usually available to command buttons are the enable, disable, visible, invisible, caption, size, picture, and click.

Labels

Labels are pieces of text that make it easier for the user to understand what is presented on the screen. Labels are not editable when the program is under execution. In Figure 9.6, a label is used: "Check all that applies." This guides the user on how to use the control. Usually each of the controls on the screen would have a label associated with it to indicate its name, function, or an explanation. Labels can be made invisible or visible. Labels are very important controls on the screen, and just changing the labels on the screen can change the software from one language to another.

In addition to the controls described earlier, the specific programming language being used may provide some more controls. One point we need to note is that some of these controls like the text box, combo box, list box, and the grid, can be used as both input as well as output.

Links

In the present day where the applications are Internet-based, links are provided to navigate the user to other pages within or without the application. Links are two components, namely, the URL (Uniform Resource Locator) executable by the browser and the text that is displayed to the user to aid him/her. Of course, the URL itself can be used as the text, but using a humanly readable text is preferred to make the application user-friendly. Sometimes, the URL could be too long, and it is preferable to use a piece of text in place of URL. Each programming language provides a facility to define links and link them with pieces of text to be seen by the end user.

Input Statements within Programs

The controls described earlier are used to provide a GUI to the user, but we still need to program the computer to process the inputs provided by the user. In Inputs, we program the computer in two aspects, namely, to validate the data received and to store the data received. First, we shall discuss the storing statements and then discuss the data validation statements.

There are usually two input statements when we are using a CUI (Character User Interface). These are:

1. To get just one character.
2. To get the value for one variable or more variables.

The C-family languages use the "getchar" statement to receive one character read into a predefined variable. Other languages have similar keywords for receiving just one character. This statement can also be used to receive large amounts of data from a file character by character. This keyword is the main tool in commands for copying files.

To receive value for a variable, the C-family languages use the "scanf" keyword. The keyword "scanf" can receive numeric as well as non-numeric values. However, we need to define the variable appropriately before reading the value from the keyboard. If there is a mismatch between the type of the defined variable and the data supplied from the keyboard, it would raise an error. Other languages use key words like "read" and "input" to receive values into variables. Of course, the type of the defined variable and the value supplied must be the same. Otherwise, an error will be thrown up.

Whatever values we enter from the keyboard in response to the previous statements, the value would be displayed on the computer screen for us to view and ensure that it is the right value. We can also correct the value if necessary by using the backspace key. While we can present a blank screen to the user to enter the input, we better provide him/her a prompt (a message on the screen that instructs the user what needs to be done) on the screen to guide him/her. We can also locate the prompt appropriately on the computer screen at a location we feel is most convenient for the user to read the prompt and enter appropriate data. The value will not be input to our program until the user presses the Enter key on the keyboard.

The longest string that we can enter into a variable is a double-precision number for numeric variables and a maximum of 255 characters for the character variables. If we wish to have the user enter large amount of textual matter, we need to write special programs like MS-Word or use database packages.

Using the scanf family of statements, we can receive input for one or more variables into our program. These statements can also be used to read data from flat files. When we use this scanf family of input statements, we need to specify the input device from which to get the input data. Usually, it would be an integer defined earlier when we opened the file for input. If we do not specify a device ID in this statement, it usually expects the input from the standard input device, which in the present day is the keyboard and screen combination. Inputting multiple variables would follow a specific set of protocol based on the specific programming language being used. When receiving input from a file using this class of input statements, there would be extensions to this statement to suit the type of data file organization of sequential, random, or indexed sequential organization. Receiving information from a database depends on the specific DBMS software we are using. However, there is a standard set of SQL constructs that are common to all DBMS software packages. However, they differ in the statements used for connecting to a database and then opening a database table. For reading, of course, the SELECT statement is used across all DBMS software packages. Here are some important standard SQL statements:

1. *SELECT*: This statement is used to read information from a database table. SELECT keyword is followed by a * (indicates reading of all the fields in the record, or in other words, the entire record) or by the list of the fields to be read. The field names must be written as defined in the database table structure. This statement can include a relational or logical expression to filter the retrieved records.

2. *INSERT*: This statement is used to add a new record to the database table. Of course, this statement also causes the update of the relevant indexes used on the table. The INSERT keyword is followed by the field names and the values (constants or variables) to be stored into those fields.

3. *UPDATE*: This statement is used to modify the contents of fields in an existing record. In this statement, we can change the contents of one field, some fields, or all the fields as necessitated by the situation. The UPDATE keyword is followed the names of the fields to be modified along with the values (constants or variables) with which to modify the contents of the fields. This statement can include a relational or logical expression to locate the records to be updated. This statement can be used to modify the contents of one record, multiple records, or all the records as necessary.

Of course, there are many other SQL keywords that aid in the programming and manipulating the data in the databases. For a comprehensive coverage, you need to refer to an SQL programming book or manual of the specific DBMS you are using. DBMS packages also make available certain arithmetic statements like count, average, sum, and so on to retrieve the data as well as perform the simple arithmetic operations.

Data Validation

"To err is human, but to make a real mess, you need a computer," or so goes the joke about data errors in the information obtained from computers. Most of the errors are attributable to errors made while entering data into computers. A few of the errors are caused because of defects in the programs. The errors caused by the defective programs are easily traceable as they are consistent over all the data that is processed by the program, but data entry errors are difficult to detect and rectify.

In the past, when the data entry was offline, the data was entered by data entry specialists and each item of data was verified by reentering the data or by manually checking to ensure that the data was as accurate as humanly possible. In those days, the data was accurate to 99.96%, or for every 10,000 records, 4 records were likely to be containing defects.

Now, offline data entry is still used but not in mainline business applications. Most of the data comes from business transactions handled every day by employees working in organizational business processes. The people entering data are not data entry specialists, but functional specialists using the computers for record-keeping and other purposes. They are not likely to enter every item of data twice to ensure the accuracy of the data entered in the computer. Therefore, we need to build in data validation statements to ensure the accuracy of the data and to prevent as many errors as possible from entering the corporate databases.

We have to note that data-entry errors caused purposefully cannot be avoided. By including data validation statements in our programs, we aid the functional specialists using the computer from committing errors inadvertently. Here are some of the methods we use in programs to prevent inadvertent errors:

1. *Masterfile look up*: We usually have some master data in our databases and, each table must have a primary key, which is an important nonduplicated item of data. In a payroll or HR application, the employee ID is a key item. In a marketing application, the customer ID is a key item. In a purchase application, the order number is a key item. In a retail store, the product codes and the prices are key items. So, whenever an item of such key is required to be entered, we force the user to select from a list using a combo box or another appropriate control. That way, the user is prevented from entering the wrong data. When the user needs to enter data for such an item, we check the appropriate database table for the existence of the item, and if it is not available, we flash an error message to alert the user about the mistake committed.

2. *Range check*: This is useful in preventing errors while entering numeric data. For example, in a HR application, we check the date of birth to be within the range of employability to ensure the employee attained the minimum age and is below the retirement age. When processing the financial applications, we see the range of permissible amounts.

3. *Logical checks*: We use certain common-sense logic to see if the entered data is accurate. For example, when the user enters the date as February 29, we check if the year is a leap year or not. In an email address, we expect one "@" character, a dot, and, at the end, there are certain acceptable words such as com, edu, org, and so on. In a website address, we expect the letters "www," and at the end, acceptable words like com and org. These are few simple examples, but in every application there will be many situations like this, and we build in as much logical checking as possible to ensure accuracy of data.

4. *Unusual data*: This is used in financial applications. When a payment to a person/ employee/organization goes above the usual amount by a certain percentage, we alert the user before processing the data. In credit card and debit card applications, we not only check the limits, but also compare the present transaction with all the transactions over the past year or so to prevent fraudulent use of the card.

5. *Unusual requests*: With the Internet making it easy to access data all over the globe, we now look at the source from where the request is coming. If the request comes from a foreign location, and especially from an enemy country, we prevent access and alert the administrator.

6. *Unexpected characters*: This has become a big problem in these Internet applications. As you perhaps know, HTML (Hyper Text Markup Language) is built up using tags. Now, if the data is entered into the text boxes with an executable string, it can cause havoc. We ensure that such tags are either prevented or prevent their being used as valid HTML statements. This has now become an important aspect of today's programming. Another aspect is, while entering data in fields such as names of personnel where digits and other non-alphabetic characters should not be present, we check for such unexpected characters and eliminate them or alert the user about the erroneous entry.

7. *Integrity check*: When codifying important IDs such as social security numbers, employee IDs, material codes, order codes, and so on, they are built with some logic. When someone enters such data, the data will be verified against the logic to ensure that accurate data is being received.

8. *Type-checking*: We also check the data being entered to ensure that right type of data is being entered. Common mistakes are entering an "O" (letter "O") in place of zero. This is especially implemented when accepting numeric data.

9. *Constraint-checking*: In many cases, we place constraints on the type of data being entered. In numeric cases, there could be a minimum value or a maximum value. In character strings, there could be a minimum length and a maximum length of the string acceptable to the application. In some cases, there needs to be only one word. We check for all such constraints. In passwords, it has become common to ask for a numeric and a special character. We do that in this kind of checks.

10. *Consistency checks*: We check for consistency between the parts of the data being entered. If the title was entered as "Mr.," then the gender should be "Male." If a product is selected, then the price and discount should be commensurate with the product. There could be many such consistency checks possible depending on the application, and we check for these aspects.

Receiving Data from Cars, Rockets, Airplanes, and Machines

In programming computers that control machines like rockets, airplanes, ships, submarines, automobiles, and a host of other machinery used in the world, the data comes from machines. How is this data gathered and transmitted to the CPU that processes it and returns appropriate instructions back to the machines?

Each machine, be it a rocket, airplane, or car, has certain parameters that need to be monitored to keep the machine running in a healthy manner and do what is expected of it. These are classified into internal parameters and external parameters. Internal parameters are:

1. Temperature in the engine or wherever heat is generated.
2. Pressure inside the places where pressure is generated.
3. Velocity/speed of the moving parts of the engine and product itself in the case of rockets, airplanes, and cars.

4. Vibrations of places at some critical places like shafts, enclosures, and panels.

5. Various measurements like the voltage, current, electrical resistance, volume of liquids, levels of cooling liquids, temperatures, and so on.

6. There could be situation-specific parameters internal to the machine.

The external parameters matter especially to moving machines like the automobiles, airplanes, ships, submarines, rockets, and so on. They are:

1. The environmental parameters like the wind speed, rain, snow, dust, and so on.

2. Any approaching objects.

3. The direction and the ground speed.

4. The time allowed, elapsed, and remaining.

5. The distance to be covered, already covered, and remaining.

6. There would be many more machine-specific parameters.

All this data comes from instruments/sensors mounted on the concerned parts of the machines, and they collect the data that is basically analog in nature. They would have either built-in or externally mounted A2D (Analog to Digital) converters that will convert the analog signal from the instrument to a digital signal suitable for a digital computer. Now this signal is interpreted and processed by the CPU embedded in the machine. As computer programmers, we are not concerned how these A2D converters work. We simply assume that the data comes in digital form readily usable by the CPU and code our programs.

Final Words

Inputs are vital in computer programming. Inputs can come from master data stored inside the computer and transaction data generated with every business transaction performed. Data comes in from various input devices including flat files and database tables. The important input programming keywords are discussed, but the specific syntax depends on the programming language selected for the application. During business transactions, data comes in from GUI screens housing various controls, and the important controls are discussed in this chapter.

10

Output Statements

Introduction

The purpose of using computers is to generate useful and actionable information by processing data. Once generated, the information needs to be delivered to the user on the desired medium so that the users, who need the information, can utilize it. In this chapter, we will discuss how the information is sent from inside the computers to the external world.

Output versus Enquiry

Some people differentiate between the output and the enquiry. Enquiry consists of an input and an output that is generated based on the input. IFPUG (International Function Point Users Group) mandates that it is referred to as enquiry only when the output part in the enquiry is just an extraction of data without subjecting it to any type of processing. IFPUG mandates if the extracted data is subjected to any processing, then we have to treat the enquiry as an input and an output. This distinction is more applicable to those sizing the software than to us, the programmers, who just deliver the functionality. For us, an input is receiving data from the users and an output is delivering processed or unprocessed data to the user.

Structure of Output Statements

Where do the outputs we generate go? Obviously, they go to some output device. These output devices could be printers, other machines, plotters, computer screens, and so on. We discussed these output devices in Chapter 1. Here, we are concerned about programming the output statements, so they go to the right output device. Generally, the output statement consists of three parts:

1. A keyword such as "write," "print," or some such other keyword that specifies that the purpose of this statement is to deliver some output.
2. The ID of the output device which indicates the desired output device that is going to receive the generated data/information.
3. The data or information that was generated by our program.

There would be other components in an output statement in other cases. We will discuss them at appropriate places.

Output to Flat Files

We have discussed flat files in Chapters 3 and 9. We have noted that the flat files can have different access modes like random-access, sequential-access, and indexed sequential-access files. These can be housed on magnetic tapes or magnetic disks, but today, magnetic tapes are used as back-up devices rather than as I/O devices. To use flat files, either for input or output, we need to first open them. What is involved in opening the flat files was described in Chapter 9, but it slightly differs in the case of opening flat files for output. First, we need to understand the modes of opening the flat files for output. While input from files is reading using the access mechanism of the desired file, output has three modes of output, namely,

1. *Add or append mode*: In this mode, the contents of the file would be retained as they were, and new records would be added to the end of the file. The file would be lengthened by the addition of the new records. In this opening statement:
   a. The file is located on the magnetic disk connected to the computer. If the file could not be located, an error message indicating that the file could not be located would be flashed on the screen.
   b. If the file is located, the details about the file, location, record structure, and number of records, etc., would be copied to a location in the RAM.
   c. Sufficient RAM would be allocated to hold a record or a block of records. This space would be used to receive the saved records.
2. *Output mode*: When a flat file is opened in output mode:
   a. If a file with that name already exists, then it would be deleted. Therefore, we need to be careful when opening a file for output mode lest we may lose an existing file.
   b. A new entry is made in the VTOC with the name specified in the file opening statement, and disk space would be allocated adhering to the procedures of the OS.
   c. Sufficient RAM would be allocated to hold a record or a block of records. This space would be used to receive the saved records.

3. *Modify/update mode*: This facility is generally not available in flat files. Flat files, by design, are not amenable to modification of contents. When a flat file is to be modified, we create a new file, taking the contents of the existing file and the changes as input. But in certain specific cases, modification of flat files is allowed. These conditions are:

   a. The record structure must contain fields of fixed length. That is, the field must contain the same number of character in every record. If a record has a lesser number of characters, it must be padded with either blank spaces or leading zeroes to fill the length available for the field.

   b. The file needs to be organized either in random-access mode or indexed sequential-access mode. While sequential-access flat files do not allow modify/update mode, certain computers may allow such facility. But it is very rare.

Whenever we issue a write statement, the record would be saved in the RAM allocated while opening the file. Whenever this space is filled, the OS would automatically write the information to the disk and empties the space to receive new records. The residual records in this RAM would be written to the disk when we issue the "Close" file statement. Of course, modern programming languages flush all the buffers at the time of terminating the program execution, but as programmers, we cannot rely on such facilities. We need to include the close-file statement to ensure that all the information received is stored in the files without exception.

Writing information to flat files consists of three steps, namely, the opening of the file, writing information into it, and then closing the file. The opening statement for a flat file would consist of the following components:

1. The keyword "Open" or an equivalent keyword to indicate to the computer that the requested operation is file opening.

2. The file ID, which is usually an integer, and in some programming languages, it is to be declared as a special type of "File." In others, it is just an integer assigned by the programmer using a file name. This file ID would be used in all output statements that send information to the file.

3. The name of the file, including the disk identification, the directory or the folder, and the name of the file with extension, if any. If there is only one disk, then we need not supply the disk ID on which the file is resident.

4. The mode of opening, namely, input, output, or modify.

5. The access mode, namely, sequential, indexed sequential, or random access.

This results in the file being opened for the specified mode. Now, we can use the file to send information using the defined file ID.

Once we complete writing data to the file, we must close the file. It is generally a simple operation. The keyword in most languages is "close," followed by the file ID. The closure action would cause the data in the buffer to be written to the file and the memory space allotted to the file operation is released to the OS. If we do not close the file using the specified statement, then the data in the buffer is likely to be lost.

The write statement consists of the following components:

1. The keyword "Write," or "Print," or some such other meaningful keyword that indicates to the computer the nature of the operation requested.
2. The file ID, to instruct the computer where to write the information.
3. The format of the output.
4. The list of variables to write.
5. Any other language-specific information.

What is the format of the output? In some programming languages, mere mention of the field names is adequate. In some programming languages, we need to specify the format of each of the variables mentioned in the output statement. The format of each field could be:

1. For numeric data—####.##, 999.99, or something like that—the # character or the digit 9 indicates a numeric position and the dot (.) indicates the position of the decimal point. Now, if the format is "##.##" and the data is 34.2, then the formatted output would be "034.20." The empty spaces to the left of the decimal point would be filled with leading zeroes to the number and the empty spaces to the right of the decimal point would be filled with lagging zeroes.
2. For alphanumeric data, the empty spaces would be filled with blank spaces. Now where would the blank spaces be filled—on the right-hand side or the left-hand side? Some programming languages, by default, fill the blank spaces on the left-hand side, and some fill the blank spaces on the right-hand side. Some programming languages expect the programmer to specify where to pad the blank spaces. This is referred to as the "justification." When we specify "left-justified," the significant data would be to the left-hand side and the blank spaces would be filled in on the right-hand side. When we specify "right-justified," the data would be placed on the right-hand side and the blank spaces would be filled in on the left-hand side.

Now what happens if the data supplied and the format differ from each other? Ideally, an error should be thrown up by the OS. But sometimes, the erroneous character in the data would be replaced with a blank space or a zero and written to the file. So, we must take care to ensure that the data and the format are exactly the same while writing the data to a file.

What would be the final shape of the record in the file once the output instruction is executed? There are two types of data records in the flat files. They are:

1. Fixed field-length records
2. Variable field-length records

In fixed field-length records, each field in every record would contain the same number of characters with the shortage being filled with either blank spaces or zeroes. In variable field-length records, the number of characters in a field can vary from record to record. Let us illustrate this with an example.

Let us assume three fields, a first name of 6 characters, a last name of 8 characters, and an income of 8 characters. Now this would appear in a fixed field-length records as this:

JOHN   MCDONALD00050000
SHANE WARNER   00100000

You can see the blank spaces being padded for the alphanumeric data and leading zeroes for the numeric data. Now the same data, in variable length records, would appear like this:

"JOHN", "MCDONALD",50000
"SHANE","WARNER",100000

Now, as you can see, there would be no padding, but the alphanumeric data is enclosed between quotation marks and there is a comma separating each field. Some programming languages do not use the quotation marks, but the field separator, generally referred to as the field delimiter, is used in all programming languages. The field delimiter can be any character but generally the comma or the semicolon are the most popular field delimiters.

How is one record delimited from the other? In fixed-field length records, they would not be delimited. The earlier two records would appear on the disk, thus:

JOHN   MCDONALD00050000SHANE WARNER   00100000

When we try to read such records using an editor, it would be confusing. But the program can distinguish clearly, as it counts the number of characters and then selects the right record. To make reading such records simpler in an editor, another file format, referred to as Line Sequential, is often used. In this format, each record is delimited using an enter (ASCII 13) character, a line-feed character (ASCII 10), or a combination of both. While using a record delimiter makes it easy for us to open the data file in a text editor and make corrections, it increases the file size on the disk.

In variable field-length records, too, sometimes a record delimiter as explained earlier is used for the same purpose of opening the data file in a text editor to make corrections to data as necessary.

How do we write programs to ensure that the output is written as either fixed field-length records or as variable field-length records? Usually there would be a different keyword in the programming language to denote the type of record. Some programming languages use the "write" keyword for fixed field-length records and the "print" keyword for fixed field-length records. Other languages use other keywords to specify the type of record to differentiate the record types.

While writing data, we need to ensure the following:

1. Some programming languages mandate that the data be moved into the record fields. Therefore, the data that resides in the variables needs to be moved to the output record fields by an assignment statement.

2. We need to ensure that the variable names appear in the write statement in the same order as they are to be stored in the flat file as defined in its record structure.

3. The fields must be the same data type as in the variables being written to the file as in the definition in the file structure.

4. The field lengths must also be same in the file structure and in the variables being written to the file.

When these precautions are ignored, errors are likely to be thrown up by the OS. We are lucky if the errors are thrown up, as it allows us an opportunity to correct the mistakes and re-execute the program. In some case, no error would be thrown up and the OS would write the information as specified in its functionality. That would result in erroneous data files.

Of course, the use of flat files for storing data has come down significantly with the growth in the use of DBMS packages. Still, flat files are used for storing configuration data in commercial applications, and flat files are the main data storage method in the real-time applications. Flat files have very little overhead in terms of storage space and, therefore, wherever there are constraints on storage space, flat files are used.

Output to Database Tables

Most of the data being stored goes into database tables in the present day, leaving a small amount of data to be stored in flat files. DBMS packages provide numerous advantages over flat files for storing data and hence are used heavily in commercial applications and where there is no pressure on the available storage space. Now let us look at how output is directed to DBMS tables.

These are the steps in sending output to database tables:

1. Establish a connection to the database on the database server.
2. Open the database table.
3. Insert a new record.
4. Update an existing record.
5. Close the database table.
6. Sever the connection established earlier to the database.

Now let us discuss each of these steps in detail.

Establish a Connection to the Database on the Database Server

In the pre-Internet days, the database and the application were on the same machine. No longer is that so. The application can be running on one machine and the database could be on another machine. To establish a connection with the desired database, we need to know the following aspects:

1. The IP address of the machine on which the DBMS was installed. The IP address can be in the standard form (such as 192.168.0.1), or a name such as chemuturi.data.
2. The name of the DBMS package—this helps in making the connection string to connect to the database. Each DBMS package needs the arrangement of information in a specific order in the connection string.
3. The name of the database.

4. The type of security used by the database. In some cases, the security could be integrated with that of the server. In this case, unless the user is able to connect to the server, the user would not be allowed to connect to the database. The security rights need not be defined separately for the user on the database. The database would use the security definition of the server machine. The other security method used is to define the user rights on the database. In this case, the user need not have rights to connect to the server machine to connect to the database. The connection string would differ in both the cases.

5. The username and password to connect to the database.

With this information in hand, we are now ready to prepare the connection string. The connection string is nothing, but the previous four aspects concatenated into a string of characters. We may use four statements using the character arithmetic for concatenating the strings, or write one single statement. I suggest writing four statements, as it becomes easier to debug. You would be amazed at the propensity to err in the programmers. I would go to the extent of saying that programmers in the original development of the programs and later on in maintaining them spend much more time in debugging the programs than in writing them. So, better to write a longer program so we can save a lot of time while debugging it. It is common to assign the connection string to a variable and use it whenever we connect to the database. Usually, we store the string along with the open statement in a subroutine that is accessible to all the programs in the application and call it whenever we need to connect to a database.

Now we simply write the statement for to connecting the database along with the connection string. Usually, an OPEN statement is used to connect to a database. The syntax is usually like this:

```
Open <ConnectionString>
```

"Open" is the keyword that tells the computer to establish a connection with the database using the parameters given in the connection string. We need to write an error-trapping statement immediately after the connection statement to trap the error should it occur. There are any number of reasons why a connection with the database cannot be established. Here are some:

1. The connection between the machine on which the application is running and the database server may be in a disconnected state.

2. The server machine may be suffering from a hardware/software malfunction and is not working as it should.

3. The number of connections on the database server have reached the maximum. As stated earlier, a single-user DBMS can accept only one connection at a time. Multiuser DBMS packages can accept multiple connections concurrently, but the price of the DBMS varies with the number of connections it can accept concurrently. So, if we have installed a 5-user DBMS and are attempting to establish a 6th connection, obviously it fails.

4. Each connection to the DBMS needs certain amount of RAM in the database server. If the RAM is full and the server is not able to allocate more RAM for another connection, or it is not able to allocate the RAM to the new connection request in the expected time, the server my return an error message.

There could be more reasons like these. Whenever a connection is not established, an error message is returned to the calling program by the database server of the machine on which the application is running. We need to trap this error message, decipher it, and flash a message to the user in a meaningful manner so that he/she is not panicked and runs to the system administrator or lodges a needless complaint.

Once we have established a connection with the database, we can open any table in that database.

Open the Database Table

To open a database table, connecting to the database is a prerequisite. Once we have connected to the database, we can open a table from that database. Different DBMS packages have different protocols. Opening a database table involves the following:

1. Defining a handle (a variable containing a numeric value) to refer to the table. This has to be unique, that is, the same variable should not be used for two tables concurrently.
2. Allocating a sufficient amount of RAM to hold one record or a block of records. Usually, RAM is required for a block of records. The block size differs from DBMS to DBMS and OS to OS.
3. Reading records from the table and loading them into the allocated RAM.

In the statement opening the database table, the following components are present:

1. A keyword that tells the computer to open the desired database table.
2. The name of the table.
3. The lock to be applied on the table. This lock is usually of three types:
   a. *Read-only*: This lock allows the current program only to read the existing records. It prevents the current program from inserting new records. It also prevents the existing records from being updated.
   b. *Read and write*: This type of lock allows the current program to read the existing records, insert new records, and to update the existing records.
   c. There could be other types of locks specific to the DBMS package. One such is to provide a situational lock. It allows the records to be read and written depending on the availability of the database. If no other connection is using the database, the current program would be allowed to write or update records. Otherwise, an error message would be returned. This is one method people use to allow a single-user database to be used by multiple users concurrently.
4. If there is security for the table besides the security on the database, we need to supply the user ID and password for the table.

Now, we form a string in the order specified by the DBMS package and issue the statement to open the database table. Once we open the table, we can extract the records from the table.

To extract records from a database table, we need two pieces of information in addition to the name of the table:

1. The condition (relational or logical expression) to filter out the records to be extracted from the table.
2. The list of fields to be extracted for each of the records. Usually, a star (*) in the space for listing the fields results in extracting all the fields in the record.

We usually use the SELECT keyword in almost all the DBMS packages. Actually, the standard SQL specifies the SELECT keyword for extracting the records from a database table. In most DBMS packages, the table opening and the record extraction statements are merged into one statement, but there could be different statements for these two actions.

Insert a New Record

There are two methods of inserting a new record. One is to insert a blank record and update each field with the data, and the second is to use the INSERT statement provided in the SQL. The INSERT statement is the preferred method to insert a new record. In flat files, we can only append a new record at the end of the file. But in database tables, we can insert a record in the middle of the table and not necessarily at the end. Physically, the record needs be located at the end of the table, but the DBMS achieves this logically by modifying the links that connect one record to another. Another advantage of using the INSERT statement is that, it initiates the process of reindexing the table automatically. Usually the syntax of the INSERT statement is written in this manner:

```
INSERT INTO <table name>
(filed1, field2, field2, … fieldn)
VALUES (value1, value2, value3, … valuen)
```

where:

Field1, field2, field3, etc. are the names of the fields defined in the table
Value1, value2, value3, etc. are the values with which the fields would be filled

Here are rules to be adhered to for using the INSERT statement:

1. The number of values supplied must be the same as the number fields mentioned.
2. The order of values supplied must be the same as that of the fields mentioned in the statement.
3. The data type must be the same for both the field and its corresponding value.
4. The values can be constants or variables, but as we noted earlier, hard-coding values in the programs is a bad practice. So, we ought to use only variables in this statement.
5. If we do not enter all the fields available in the table, the remaining fields will be filled with NULL values or the default values defined in the table design.
6. The primary key field must be included in the list of fields unless it is an auto-incrementing field.

7. In the case that a record with the same primary key already exists in the table, an error is thrown up by the DBMS. There is an UPDATE statement to modify the contents of an existing record.

8. We need to write an error-trapping statement immediately after an INSERT statement to ensure that the record is properly inserted.

In some DBMS packages, it is possible to insert a blank record and move data into fields using assignment statements. The blank record is appended at the end of the table and after the primary key field is entered, it will be indexed. The advantage in this method is that it is very easy to debug when the insertion fails. In the INSERT statement, all fields and their values are in one statement, and sometimes it becomes very difficult to locate the troublemaking value. In the second method, as we use a separate assignment statement for each field, we can easily locate the value that is causing the trouble and correct it. So, in my humble opinion, it is better to use the second method of inserting a blank record and fill the fields with values, if the selected DBMS package allows it.

Modify the Contents of an Existing Record

Modification of data is also a kind of output statement, and we need to modify data in the existing records more frequently than we insert new records. We have two methods for modifying the data in records. The first one is the UPDATE statement provided by SQL, and the other is to locate the desired record and move data into the fields needing modification using assignment statements. The UPDATE statement usually takes the form:

```
UPDATE <table name>
SET (field1 WITH value1, field2 WITH value2, field3 WITH value3 … fieldn
WITH valuen)
WHERE <logical expression>
```

where:

- Field1, field2, field3 … fieldn are the field names as defined in the table.
- Value1, value2, value3 … valuen are the values of the data that go into the fields.
- Logical expression used to locate the desired record.

Here are rules to be adhered to for using the UPDATE statement:

1. The data type must be the same for the field and its value.
2. The values can be constants or variables, but as we noted earlier, hard-coding values in the programs is a bad practice. So, we ought to use only variables in this statement.
3. We need to write an error-trapping statement immediately after an UPDATE statement to ensure that the record is properly updated.

One overwhelming advantage of the UPDATE statement is that we can modify multiple records with one single statement. We made use of this facility during the Y2K conversion days to update the tables for changing the 2-digit year to a 4-digit year with one statement per table. It saved significant amount of time for the organization and reduced the tedium for the database specialists.

In some DBMS packages, it is possible to locate the desired record and move data into the fields using assignment statements. The advantage in this method is that it is very easy to debug when the update action fails. In the UPDATE statement, all fields and their values are in one statement, and sometimes it becomes very difficult to locate the troublemaking value. In the second method, as we use a separate assignment statement for each field, we can easily locate the value that is causing the trouble and correct it. So, in my humble opinion, it is better to use the second method of updating single records and modify the fields with values, provided the selected DBMS package allows it. But if multiple records need modification with the same values, then the UPDATE statement is the best fit.

Output to Files Like Excel, Word, PDF, and So On

We may need to send the output of processing to files like MS-Excel, MS-Word, MS-Project, PDF, and such other files. How do we do it? There are no general statements that are common to multiple development environments. Each development platform provides tools in their SDK (Software Development Kit) to achieve this kind of output. Usually, the following steps are part of sending output to these kind of files:

1. MS-Excel and other spreadsheets are to be treated like a DBMS package.
    a. We need to open a connection to the desired spreadsheet file.
    b. Open the spreadsheet file.
    c. Select the worksheet. If we do not select a spreadsheet, the output will go by default to the first spreadsheet (or worksheet).
    d. Move data into each cell using assignment statements of a row.
    e. Using a loop, fill all the rows with the output data.
    f. Close the spreadsheet file.
    g. Close the connection.
    h. Just as in the case of database tables, the spreadsheet file must be existing to be connected and opened. Without an existing file, we cannot establish a connection. If we wish to create file, we can do it using system calls to the OS to create the file. But to establish a connection, we need to have an existing file.
2. MS-Project is very similar to the MS-Excel except that it has predefined columns. The steps in delivering output to an MS-Project or a similar software package are same as described in the earlier bullet on delivering output to MS-Excel or other spreadsheet files.
3. Delivering output to word processing files such as MS-Word.
    a. We need to create a file using system calls to the OS to create a desired type of file, if it is not already existing.
    b. Establish a connection to it using the facilities given in the development platform. It would be similar to establishing a connection to the database, but there would be differences of keywords.
    c. Word-processing files are just streams of characters with tags for features like bold-facing, italics, links, and so on. We need to deliver these special tags also along with the textual matter. How to send these would have to be learned from the usage manual of the development platform or from its help pages.
    d. Word-processing packages accept simple textual matter, and we can simply create a text file and give it the desired file extension and it can be opened by that package, but the embellishments have to embedded by the user.

4. PDF files are also streams of characters, but they have a header and a footer. We need to add the header and footer to each output before they are accepted as PDF files. Creating a text file and giving it the PDF extension will not serve the purpose. Of course, we also need to embed tags for text effects along with the text to achieve the embellishments on the text or diagrams. Alternatively, utilities are available to convert textual matter into PDF documents. We can call them in our programs and direct the output through them to create PDF files.

5. There could be other types of files that would require us to send outputs to them. We need to study the requirements of each file, study the usage guide of the development platform, learn the series of commands and tags that are essential in sending output to these files, and implement them in our programs.

Sending Output to Machines

Now, we are using computers for controlling various machines ranging from washing machines to rockets. There are certain commonalities in sending outputs to machines. Then there are specifics for each machine. Let us first look at the commonalities and understand how to send outputs to machines.

1. When we send outputs to machines, the instruments designated for interfacing with the computer receive the information. Instruments convert the received data into commands and pass it on to other instruments designated for controlling the machines.

2. The instruments are connected to the computer by a special port on the computer. Usually, the port follows the protocol defined in the standard 802.3 of IEEE. We need to learn its protocol for gaining the ability to send outputs to instruments on machines. Of course, some machines may use their proprietary protocols. We need to learn them when faced with such a situation.

3. We usually receive a driver software program with each machine. The driver software provides guidelines on how to communicate with the machine. We need to call the routines provided by the driver software to send outputs to the machine.

4. Of course, we need to install this driver software utility on our computer as well as the target computer for our programs to work.

Then each machine, depending on its class, would have some common functionality. For example, all cars have to move, brake, accelerate, and so on. Each washing machine will have wash cycle, rinse cycle, start, stop, warnings, and so on. These will be common to a class of machines but specific to every model or make. We need to learn these specifics for each occasion and code our programs suitably.

One very important aspect we need to understand to be able to write programs to control machines is that these programs rely heavily on using system calls. A "system call" is a programming statement in which we issue a command to a machine using the facility provided by the development platform to pass control to another program and receive control or feedback from that program. In fact, issuing SQL statements for managing database tables are also system calls. Each development platform provides keywords for using system calls. We need to understand these keywords thoroughly. The following process takes place when using system calls:

1. When the computer encounters a system call during the course of program execution, it places the program execution in wait state and passes on the command to the designated machine.

2. The machine receives the input from the computer through its communication hardware and initiates its internal program execution.

3. The machine executes that initiated program and performs the ordered functions.

4. The machine sends back the result of execution to the computer. The result could be:

    a. Signal to indicate that the program was successfully executed. In this case, the program activates the concerned program in wait state and puts it in ready state so the computer can execute the next instruction in the program.

    b. Signal that the program execution failed with an error code. Now the computer decodes the error code and hands over control to the subprogram that handles errors.

    c. Signal with the information requested by the computer. In this case, the computer passes the received result to the waiting programs and puts it back in ready state for execution.

This is how a system call is handled by the computer. So, when attempting to write machine-control programs, we need to master the methodology of coding system calls and also learn the programming commands provided by the machine.

Output to the Screens

Usually in most development platforms, the screen is the default output device. While writing programs, we need to specify the device on to which the output needs to be delivered. If we do not specify any device, the computer sends the output to the default output device, which is usually the screen. This output is displayed at the position of the cursor. Various output statements are available in different programming languages, such as:

1. Write
2. Print
3. Printf
4. Display
5. And so on

Usually the syntax of these statements would be like this:

```
Printf (variables, constants \n)
```

where:

- Variables are data items
- Constants are hard-coded values
- "\n" indicates printing a line-feed character or moving the cursor to the next line

We can also format the output as we need to. Other output statements have different syntax. We have to learn the syntax rules of these output statements in the programming language selected for our project. These statements are now mostly used to deliver outputs to the printer when we do not use a report-generation utility. We used them and, until the GUI came along, they were the only choice. But now, with GUI, we just assign the output to controls on the screen using assignment statements.

We need to deliver the output of processing on many occasions. In fact, we deliver output to the screen much more than to a printer or a hard copy in commercial environments. Sometimes, we do both. In this section, let us see how to write programs to send output to a screen.

Bulk Outputs to Screen

It is a good programming practice to send bulk outputs to the screen first before sending them to hard-copy devices. The user can see if the processing produced the desired results with the expected accuracy and precision. That way, paper and printer ink would be saved which helps the environment, if not money! For generating bulk outputs, we use a report-generation utility which usually sends the output to the screen, giving facilities to the user to save the output as a file in various popular formats or to send it to the printer. If such a facility is not available in our development platform, we need to format the output ourselves.

Bulk data is usually found in a tabulated manner. So, obviously, the first choice is to use a control that mimics a table. Some development platforms call such a control a grid or list or something similar. The term "table" is not generally used, as it refers more to a database table than to a screen control. When we have such a control in our SDK, what we have to do is:

1. Locate the control on the screen at an appropriate place on the screen, preferably in the middle of the screen.

2. Provide adequate width for the control but see that horizontal scrolling is avoided if possible. If horizontal scrolling becomes inevitable, see that the size of the control is kept as close to the screen size as possible.

3. Provide adequate width for each of the columns. The guideline for this is the width of the column needs to be equal to the size of the data that is assigned to it.

4. If the control provides a facility to keep some space between adjacent columns, provide a gap of half-a-character between adjacent columns. If such a facility is not available in the control, then we need to provide a gap of one character between adjacent columns. This gap will help the users in distinguishing the data easily in adjacent columns without getting confused.

5. We need to provide control statistics on the last/first screen to help the users in ascertaining the efficacy of data processing. Control statistics are explained in the subsequent sections of this chapter.

6. Alphanumeric data has to be left-justified in the column and numeric data has to be right-justified. Always include at least two digits after the decimal point in numeric values. All column headings have to be justified the same way the data in columns is.

When such a control is not available, we need to format the data on the screen using programming statements. We need to use a loop to display the data on the screen and embed the formatting statements within the loop. Of course, present day SDK's are providing some sort of control that facilitates the display of bulk data in columns. Here are the statements that help us in formatting the output:

1. *Locate on screen*: Along with this keyword, we need to supply the row number and the column number in characters. Now screens are coming in various sizes, but we need to take a screen size that is most popular at the moment and program our output. We need to use a variable for the row number so that it can be incremented in every iteration of the loop. For the column number, we can use a constant. This is one occasion, perhaps, where we can hard-code the values without being accused of bad programming! But it is a better practice to use a variable for the first column and add a number to this variable for the subsequent columns like col_loc+4, col_loc+9, col_loc+15, and so on.

2. Use a loop to retrieve data and position it on the screen using the same coordinates as in the column headings.

3. Count the number of records displayed and when it reaches one less than the number of rows that can be accommodated on the screen, hold the display and flash a message such as "Press any key to continue…." This will allow the user to read the rows displayed and see if the desired information is available in the displayed records.

4. Alternately, we may place navigation buttons at the bottom so the user can go forward or backward and also to the first record or the last record as she/he desires. This has become the standard practice today.

5. Based on the user's choice, perform the desired action, displaying the desired records until the user chooses to stop browsing the data and chooses to take other available choice.

Output to Enquiries

Enquiries are a very common action needed by users. For example, in a ticket-booking scenario, the user enquires if a ticket is available for purchase. For this, we first present him some menu of choices and after the user selects from the available choices, we retrieve the desired information and present it on the screen. This information may even be bulky, needing us to display it in multiple screens one after the other as detailed in the previous section.

If the desired information is just one item, we display it using controls like the text box, combo box, grid, and any such suitable control by moving the retrieved information into those controls.

Usually, organizations have software designers performing the design and we programmers need to implement it. Positioning of controls and the logic are given by the designers. We need to convert that logic into programming statements, so the functionality is achieved. Some organizations use graphics designers to provide the screen layout and we just have to write programs for the controls provided on the screen so that the desired actions are performed flawlessly.

Output to Printers

In the earlier days of batch processing, the printer was the default output device. In those days, we were formatting the output and sending it to the printer. We, the programmers, were controlling how exactly the output appears on the printer. Then utilities and tools were developed to make the job of delivering output to the printer easier for us programmers.

We now have report-generation tools that allow us to design the report on the screen and then call it in our programs. Let us first understand the contents of the report first:

1. *Page heading*: This would appear at the top of every page. While there is no restriction on the number of lines we can occupy for page heading, we restrict it to a maximum of four lines. If we use more lines for this purpose, we would be left with less space to place our data.

    a. The first line would contain the name of the organization. If we are using a stationary that has the logo of the organization or the organization's name as a watermark at some place, then we can eliminate this line.

    b. The second line can optionally have the department name. This is not used unless there are many departments in the organization.

    c. The third line would have the name of the report along with the date of generation and page number. This line is mandatory.

    d. We usually have one more line with the date of the report data, the page number, and the total number of pages. The date is usually at the top of the page on the same line as that of the report name at either at the left-hand corner or the right-hand corner. The page numbers would typically be at the bottom of the page on the right-hand corner.

2. *Column headings*: If we are presenting data in a tabular manner, we need to have column headings. We generally allocate three lines for this purpose, but we may need to use four lines sometimes.

    a. The first and the last lines of the column headings shall be dashed-lines to demarcate the headings from the data.

    b. We place the column headings between these two lines. We need to restrict the size of each column heading to the width provided for the data in that column. In some case, the heading may be longer than the data; if so, we need to abbreviate, if possible. If it is not possible to abbreviate the column heading, we need to put it in two lines, one below the other, taking away one more line from the data area.

    c. Sometimes, we may accommodate more than one data item in the column. For example, it is common to place name and address in one column, especially when there is a space crunch. In such cases, we need to use more lines for column headings. But we need to minimize the number of lines used for columns so we can allocate more lines for the data area. While the column headings need to explain the contents of the column, they need to be brief to conserve space.

3. *Body of the report*: In this place, we present the data retrieved and processed as necessary. We need to ensure the following when presenting the data.

    a. The alphanumeric data needs to be aligned with the left side of the column.

    b. The numeric data needs to be aligned with the right side of the column. In presenting numeric data, we need to maintain uniformity in respect of the decimal point. We need to either use the decimal point in all cases or do not use the decimal point at all. We should not present the decimal point in some cases and leave it out in others. The number of digits after the decimal point also needs to be same for all values presented in the column. If the digits in some cases are less than others, then we need to pad the blank spaces with zeroes positioned appropriately.

   c. If any value is longer than the width provided for the data, we need to wrap
       it around to the next line rather than either truncate it or occupy the adjacent
       column space. We need to check the size of the data for a size error and trap it
       to protect integrity of the report.

4. *Page totals*: It is customary to provide page totals of all numeric columns at the
   bottom of the page except for such numeric values as serial numbers, identification numbers, or codes. We provide page totals of all significant numeric values
   presented on the report at the bottom of the page. It will be in three lines with the
   first and the last lines being dashed lines to demarcate the data and the middle
   line containing the page totals. The page totals may be longer than the space provided, and we need to either provide adequate column width or wrap the text. In
   some cases, I have seen one extra line used for page totals, or totals presented in
   alternative lines to decongest and avoid size problem of the derived values.

5. *Running totals*: Running totals are not very essential, but in some important applications involving large sums of money, running totals are used. A running total
   is the sum of the data presented on the report from the first page until the present
   page. On the first page, the page total and the running total will be same. On the
   second page, the running total will be the sum of the previous page total and the
   present page total. Usually, the report-generation tool provides this facility. We
   usually use the bottom line of page totals, then present the running totals, and
   then a bottom line is used to demarcate the running totals.

6. *Grand totals*: We present the grand total at the bottom of the last page. The grand
   total is the sum of all the data presented from first page to the last page. On the last
   page, the running total and grand total would be equal. Of course, when we present running totals, we need not present the grand total. If we are not presenting
   running totals, we need to present the grand total. We use three lines or four lines
   for this purpose. The first and last line will be demarcation lines and the middle
   one or two lines are used for presenting the totals.

Report-generation tools allow us to define all these and connect the report to the database
tables, as well as set relations between the selected tables. Then they also allow us to see a pre-
view of the report as it would appear. These tools also allow us to define a subreport (a report
within a report) so that we can generate really complex reports on the fly. When executed,
the report-generation tool produces the report and displays it on the screen. It will enable the
user to save the report in a file, and it provides facility to save the report in most popular file
formats including MS-Excel, PDF, Word, and so on. As programmers, we need to master the
report-generation tool selected for use in our organization. It saves significant programming
effort as well as the tedium of programming and testing the required reports. These tools also
allow the report to be printed on the printer connected to the computer or the network.

   If we are using a report-generation tool, we need to include routines that call the report-
generation tool, along with the name of the report we designed, in our programs to generate the desired report with all the features described earlier.

Control Statistics

We need to present control statistics in all reports presenting bulk data. First, what is bulk
data? The opinion differs but in my humble opinion, if we are extracting data from a large
table and presenting most of it in the report, I would say it is bulk data. To be called

bulk data, the report must span more than three pages at a minimum. If we are counting records or lines in the report, in my humble opinion, 500 records or more can be called bulk data. It would be ridiculous to classify a two-page report as presenting bulk data. Control statistics help the users in ascertaining the efficacy of the processing. We present the following data as control statistics:

1. *Total number of records processed*: If we are extracting information from a single table, this information is easy to obtain. If we are using multiple tables, determining this value becomes dicey. In my humble opinion, it is better to present the number of records in the main master table on which other tables depend. Alternatively, we may present each table name and the number records in that table for all tables considered in the report. This is going too far, but if the customer demands, we need to provide this data.

2. *Number of records included*: We need to count the number of records included in the report and present this value. This value would be less than or equal to the total number of processed records. We may use the same rule we followed for deriving the total number of records processed for this value also.

3. We also present grand totals of significant numeric values, such as total money received or paid, the balance, and so on as applicable for the application at hand.

4. We may also present any other values specific to the functionality at hand based on the need of the users.

5. In fact, we may present the summary of all significant numeric values presented in the report.

Usually, we output the control statistics on a separate page at the end or at the beginning of the report.

Output onto Internet

Each SDK provides constructs to access and send output onto the Internet. We need to provide the following information along with the construct used for sending information on the Internet:

1. To address: This is usually the IP address of the destination to which we are transmitting information.

2. Type of information: We need to provide the type of information being transmitted by us. It could be a simple text message, a graphic, a file, an email, or something else.

3. If it is an email, we need to provide the email ID from which we are transmitting the message, the subject, and the additional IDs if we are sending a cc (carbon copy) or bcc (blind carbon copy).

4. If we are sending a database connection string, we need to provide the string with all details like the IP address, the type of security applicable, the user ID, the password, the name of the database, and any other details specific to the situation.

We need to use the construct provided by the SDK and provide the required information as detailed earlier in the specified format along with the construct. We need to provide error-trapping routines after this statement, as errors, like not being connected to the

Internet, can hinder the transmission. We also need to capture the successful transmission message and flash it back to the user.

Sending information on the Internet is an involved subject but fortunately, the SDK takes care of all the backend stuff and all we need to do is to utilize the constructs provided by the SDK and code the error-trapping routine as well as to capture the transmission information so we can inform the user with a suitable message.

Sending Information over Other Networks

Besides the Internet, there are many other networks including SPX/IPX, SNMP, SNA, and wireless networks. Each would follow its own protocol. Programming for communication would have to adhere to its programming primitives. But the following actions need to be programmed:

1. Ensure that the network is functional and that we are connected to it.

2. Send signals to the server/client machine to ensure that it is ready to receive the message. If it is not ready, then wait for some time and resend the signal, asking for permission to transmit. If the permission to transmit is not received a preset number of times, abort and flash the error message as designed.

3. Send the message in packets and receive acknowledgement that it is received properly and resend it if it was not received or not received properly.

4. Once all data is transmitted and acknowledgements are received, terminate transmission and flash a successful transmission completion message.

Final Words

Output is the most important part of data processing, as the very purpose of data processing is to deliver outputs. In this chapter, we discussed how to program delivering outputs to most frequently used output devices. Now what all we need to do is to learn the specific constructs provided by the programming language and build our program to deliver the outputs to the desired device.

11

Other Statements

Introduction

While the input statements, output statements, processing statements, and control statements form the most important sets of statements, we have many other functions to be performed by programs. These statement classes are:

1. Documentation statements
2. Starting and ending statement
3. Declaration statements
4. System calls
5. Inter-program communication
6. Interrupt handling
7. Device handling statements
8. Conversion from numbers to words

Let us discuss these classes of statements in this chapter.

Documentation Statements

Why are documentation statements needed inside computer programs? They belong in user manuals—don't they? This is a relevant question. Inside computer programs, we do not document how to operate the program, but one fact we all have to understand and come to terms with is that any good program that performs useful functions and is put into production will certainly need maintenance. By the term maintenance, I mean corrections that need to be implemented in the program due to changes in the business or technical environment. The tax rules keep changing; the technology keeps getting upgraded; the business scenario keeps changing; and in this manner, there will be so many changes that keep happening, and our programs must be updated periodically. You may be surprised that many programs written in the 1970s are still in production serving their users effectively. Even a major event like the Y2K (Year 2000) phenomenon could not replace these programs! While no accurate figures were available on the amount of money spent on Y2K maintenance, we can safely conclude that about

5–6 billion Dollars was spent by the USA alone! The cost was huge because most of the programs did not have the proper documentation to assist in easily upgrading the programs. Of course, in-line documentation was made a part of good programming standards way back in late 1970s. It is now unthinkable to write program code without documenting statements. We write documentation statements in programs in the following places:

1. *Program header*: The program header is placed at the beginning of the program and would consist of:
    a. The name of the organization for which the program was developed and a statement indicating that the intellectual property rights belong to that organization
    b. The author who initially developed that program
    c. The date on which the program was originally written
    d. The function of the program briefly in one or two lines
2. Every time maintenance is carried out on the program, we document the maintenance history as an extension of the program header, consisting of:
    a. The date of maintenance
    b. The author who carried out the maintenance
    c. The details of the software change, in brief
    d. The location (line numbers) inside the program
    e. The reference to the maintenance request that initiated the change
3. At the beginning of every set of control statements, we document the action performed and the result to be achieved by the set of control statements.
4. At the beginning of every set of looping statements, we document the purpose of the loop and how it gets terminated.
5. Whenever we use file or table open statements, we document the expected contents of the file/table to be opened.
6. If we use any complicated set of processing statements and long equations, we document the functionality briefly.

We may use documenting statements as necessary in addition to those described earlier depending on the situation at hand. Usually good programming practices mandate that the documentation statements would be 35%–50% of the programming statements. Here are some statements used in programming languages:

In the family of C programming languages like C, C++, and Java, we use the following documentation statement. We write "/*" at the beginning of the documentation statement and "*/" at the end of the statement. In the C family, we can write multi-line documentation statements using these characters. The C family refers to these statements as Commenting Statements.

In the Visual Basic family of programming languages, we use the apostrophe (') character at the beginning of the line to indicate that the line is a documentation line and is to be ignored by the compiler. In this family, we can write only single line documentation statements. If we need multi-line documentation statements, we need to prefix every line with the apostrophe character.

Documentation statements are ignored by the compiler. These statements would not be compiled into the executable program. They are there only to assist the software maintenance personnel. Documentation statements are also referred to as commenting statements.

Starting and Ending Statements

Why should we use specific keywords to indicate the program beginning and termination? Can the computer not take the first statement as the beginning and the last statement as the terminating statement? Basically, these questions are asking the purpose of the starting and ending statements. Here is the purpose of these statements.

We use the beginning statement because:

1. We have different types of programs like the main program, the subprogram, the function, the method, and so on. We use the first statement to indicate the type of program to the computer. With the help of this statement, the computer can accord appropriate treatment to the program.
2. In some programming languages, we place some code even before the first statement like header files. In such cases, this statement tells the computer to begin the program execution from here on.

In COBOL, the keyword IDENTIFICATION DIVISION is treated as the first statement. In the C family languages, the statement "main(void)" is used as the first statement of a program. Others have different keywords. Some languages like BASIC did not have any program beginning statement.

A starting statement tells the computer to begin execution of the program, that all resources like RAM are allocated, and that the process is in ready state. Then the CPU begins fetching the instructions and begins the execution of the program.

The program can end with the last statement, but it has been the practice to write a statement to indicate that the program execution is completed. This statement has special significance as follows:

1. It tells the computer to stop executing the program.
2. It tells the computer to release resources like RAM and I/O devices.
3. The OS writes any file buffers into the files and closes all the files.
4. It terminates any open database connections.
5. It removes the program reference from all places in the RAM wherever it is referred.
6. It removes the program from the list of processes under execution.

COBOL used the phrase STOP RUN as the program termination statement. The C language family uses the closing curly brace ("}") as the terminating statement for the program. Most of the programming languages use "Return" as the terminating statement for the subprograms. The actual termination statement may differ between different programming languages, but every programming language mandates a program termination statement to communicate to the OS that there are no more statements to be executed in this program.

Declaration Statements

When we write programs, we use predefined set of keywords defined by the suppliers of the SDK. This set of predefined keywords tells the computer what action is to be taken. These are like verbs in the parts of speech. These need to be followed by the objects on which to take the specified action. We add our own words to supply data to the action keywords supplied as part of SDK. These could be variables, constants, and calls to other programs. The predefined keywords supplied as part of SDK are translated into computer instructions (generally referred to as opcodes) into binary numbers during the compilation process. Then the words supplied by us are translated as addresses in the RAM where data can be found. If the word supplied by us is a call to another program, the OS resolves it as another program and takes appropriate action.

But how does the compiler know what words are valid among all the words in the program? While compiling the program, the compiler searches the set of keywords supplied as part of the SDK first and resolves them. Then, for the other words, it searches the library of routines supplied along with the SDK and resolves those words. Then it searches the library of routines we added as part of the project and resolves some words. The remaining words need to be declared within the program. If any word is not found in any of them, the compiler throws up an error.

Declaration statements aid us in declaring those words so that compiler can resolve them and prepare the executable code. Declaration statements help us in declaring the data, names of subprograms, functions, and so on. Some declaration statements are implicit. In the FORTRAN language, the letters I, J, K, L, M, and N denote integers, of which I and J were extensively used. They were so extensively used that whenever we needed a counter, we automatically declare and used i! In BASIC, any word with the suffix of "$" denoted a string, and all other words are denoted as single-precision floating point numbers. Even in present day programming languages, the word "Sub" declares a subprogram. In the explicit declaration statements, we declare data variables, arrays, files, database connection strings, table names, and so on.

In Visual Basic family, we use the keyword DIM to declare variables. Then we supplement this keyword with other keywords like Integer to indicate the integer type of data; Double for double-precision floating point number; Date for declaring a variable as a date data type; String to declare a word or a string of characters; and so on.

In the C language family, we use data-type keywords as the declaration of the data. We use the keywords like int for declaring an integer-type variable; string for a string of characters for names; Boolean for Boolean-type variable; double for a doubleprecision numeric data; and so on.

When we declare a variable, a corresponding amount of RAM is allocated to accommodate the size declared and is referred to by the name we used in the declaration statement. A value of NULL is moved into that location. Therefore, we have to explicitly move an initial value into that variable space so that we do not encounter defects during program execution due to oversight. This process of moving an initial value into the declared variable is called "initialization." In fact, many defects encountered in the field during production are due to non-initialization of variables. Most languages allow the declaration of variables to be clubbed with the initialization. For example, in the C language family, the declaration statement and the initialization statement would be clubbed and written thus:

```
Int basic_pay = 0;
```

This statement tells the computer:

1. This is a declaration statement.
2. It declares a variable by name "basic_pay".
3. The type of the declared variable is integer.
4. Its value is to be set to zero.

Most languages allow initializing multiple variables in one single statement, but one prerequisite is that they all must be of the same type. For example, here is a sample statement declaring and initializing a set of integers:

```
Int i= j= k= l= 0
```

We need to learn the different keywords provided by the SDK for declaring data and program names and then declare the required items as required by our program. Here are a few common rules to be adhered to while using declaration statements:

1. The syntax rules specified for the declaration statement must be adhered to without exception.
2. The name of the variable or data name being declared would have some rules such as:
   a. The length of the variable name is restricted. In some programming languages, it is restricted to 8 characters; in some it is allowed up to 40 characters long.
   b. It must contain no spaces inside the name, that is, it has to be one word.
   c. It permits alphabets, numerals, and one or two special characters like the underscore ("_") or the dash ("-") characters. No other characters are usually permitted.
   d. In most cases, a variable name must begin with an alphabet.
   e. There would be other rules specific to the programming language, and we need to adhere to them while defining our own words for declaration.
3. Some programming languages allow declaration statements at any point inside the program, but it is a good programming practice to place all the declaration statements at the beginning of the program. That way, it becomes feasible to assess the amount of RAM being used by the program easily without scrolling through the entire program.

Variables are classified as two types:

1. *Local variables*: Local variables are local to the program in which they are declared. The RAM allocated to such variables is released immediately upon closure of that program. It is especially useful in subprograms. The variables declared in subprograms are closed and their RAM is released as soon as the subprogram completes execution. That way, we do not hold on to finite RAM unnecessarily. These variables can be accessed only by the program in which they are declared. These variables are also referred to as dynamic variables.
2. *Global variables*: Global variables, as the name suggests, are accessible to all the subprograms and the main program. Usually all the variables declared in the main

program are global variables and are available to all subprograms in the program. They will be released only when the main program completes execution and is closed by the OS. Global variables are also referred to as static variables. When a subprogram needs to return some values to the main program, it makes better sense to use global variables than declare local variables in the subprograms, as we would be using double the amount of RAM for the duration of the execution of the subprogram.

As you can see, each type has its own place. My humble suggestion is to use local variables for processing work inside the subprogram but use global variables for exchange of information with main program.

System Calls

We write programs in higher-level languages. The first level at which the computer understands instructions is in binary numbers, that is, zeroes and ones. The next level is the assembly language of the CPU, and then we have programming languages. The programs we write are translated into machine language by the compiler. The programming languages make use of the facilities provided by the operating system, but not all of the facilities provided by the OS are used by the programming languages. For example, if we wish to read the directory listing to see if a certain file exists, we may not have the keyword to do that. Similarly, creation of a new file on the disk, read all the devices connected to the computer, and so on, are keywords that are usually not available in the programming languages. Therefore, most programming languages provide a facility to call the utilities provided by the OS or other utilities loaded on the computer and execute them from within our program and process the results obtained from such execution.

This facility is referred to as a system call. System calls enable us to call a program resident on our computer and execute it from within our program. It also facilitates us to read from various blocks of memory, such as the processes in execution, the database connections, the active process, and so on, and use that information in our program. Of course, to read such information, the security privileges must be available to the user executing the program.

Various keywords are provided by different programming languages to use system calls. In the C language family, the keyword "system" is used to make calls to the system. The command to make a call to the system looks something like this:

```
System(command);
```

Where "command" is a variable containing is a string of characters that would be a valid command of the OS. Before issuing this statement, we need to declare the variable and then move the valid string into that variable.

The output of system calls is delivered to the device determined by the command. We can divert it to a device of our choice using other extensions of the programming language.

API Programming

APIs (Application Programming Interfaces) are like system calls. While system calls are used to call programs that are on the same computer and OS, APIs are used to make calls to programs that are not of this OS and may not even be on this computer. APIs are what make the applications like Google Maps work on different computers, including mobile phones. An API is a set of subprograms, protocols, and software utilities. We can use these in our programs to call other software like Google Maps, YouTube videos, and Flickr photos and make them available to our users.

To write programs using APIs, we need to read the manual of the specific API and then issue appropriate statements in our program. However, we need to supply the following information when calling APIs in our program:

1. The name of the API.
2. The location of the API, if it is not part of our SDK.
3. The action required to be performed by the software being called.
4. The parameters needed by the called program to deliver the desired functionality.
5. The details of the protocol necessary to call the API.
6. Any other API-specific information and the information mandated by our SDK.

Once we provide this information and issue the statement in our program, it will be compiled, and the desired action will be achieved during the execution of our program. We need to study the programming manual of the specific API we are going to use before programming the API. We need to learn the keywords it provides for calling it from our programs and then implement them in our programs as required.

Inter-Program Communication

This is an important topic deserving full treatment and I have dedicated Chapter 13 for this topic.

Interrupt Handling

An interrupt is a high-priority instruction to the processor either from a hardware device or a software program. When such an interrupt is placed on the CPU, it temporarily suspends the execution of the present program and attends to the interrupt received. Actually, the CPU executes the current instruction and instead of fetching and executing the next instruction, it attends to the interrupt. An interrupt is usually placed on the CPU by hardware devices, but software also places interrupts, especially when an error that is not

TABLE 11.1

Interesting Interrupts of PC

Interrupt Number	Interrupt Description	Interrupt Number	Interrupt Description
3	Arithmetic overflow	20	Serial port interrupt
9	Keyboard interrupt	28	Ctrl+C
13	Parallel port interrupt	39	TSR (terminate and stay resident)
14	Floppy drive interrupt		

trapped by the program is encountered. The divide-by-zero error is one good example of such a software error that places an interrupt on the CPU.

For example, when we type something on the keyboard, an interrupt is placed on the CPU. If we attempt to read from the disk, the disk places an interrupt on the CPU when it is ready to receive or send information. All such interrupts are programmed by the developers of the system software.

Two terms are associated with interrupt. One is the program that needs to be executed when an interrupt is placed on the CPU. It is referred to as the "interrupt handler," or ISR (Interrupt Service Routine). The second term is the "interrupt vector," which is the reference to the respective ISR. All the device drivers are programmed using interrupt programming.

We need to develop ISRs in low-level languages or the C family languages which provide constructs to handle interrupts. High-level business-oriented languages like COBOL usually do not provide interrupt programming facilities.

Interrupt programming is like the usual programming but is focused on device handling and hardware control. To be able to develop ISRs, we need to study the programming manual of the device and then develop the needed ISR. Some interesting interrupts of PC are given in Table 11.1.

Device Handling Statements

A device driver is a software utility that is resident on your computer that we can use to receive input from or deliver output to a device. We need device drivers for all printers, plotters, DVD or CD drives, and all other such devices connected to the computer. The OS would maintain a table of all the devices connected to the computer in its RAM and would begin executing it whenever the device needs to be used. Usually, the device places an interrupt on the CPU and the CPU searches this table to retrieve the corresponding program and begins executing it.

Every device would have its own specific set of functionalities. Each device functionality would have an associated command. But this set of commands would not be the same as that of the computer. Therefore, a device driver acts as a translator, receiving the command from the computer and translating it to the device command and then passes it on to the device for execution. In addition to these commands, there will be other functionalities such as the state of the device (ready, power off, error, paper out, busy, buffer full, and so on), communication commands (ready to receive, ready to transmit, acknowledge receipt, and so on), and handshake commands (in session, session ended, and so on). A device driver program needs to handle all these commands. It also needs to handle all error conditions generated by the device. Device driver programs need to be written in low-level languages or in the C family, which provides interrupt handling and device programming.

Final Words

This chapter is devoted to explaining the concepts of using advanced aspects of programming. This is not the usual everyday stuff. These programs are required only occasionally. We attempt these programs only when we have gained reasonable expertise in a programming language and in writing software in general. We need to exercise care while developing the code as well as in testing it. The actual implementation of these aspects differs from OS to OS and one programming language to another. And all programming languages have not implemented all these aspects; therefore, when you are faced with the need to use these concepts in your programming, please thoroughly read the programming language manual, understand it fully, experiment, and only then implement.

12

Error Handling

Errors, Defects, Faults, and Failure

To err is human. We human beings commit errors. IEEE, in their standard IEEE-619 Glossary of Software Engineering Terminology, defines error in the context of programming as "an incorrect step, process, or data definition," and as an "incorrect result." Merriam Webster's dictionary, among its multiple definitions of the term, defines, for our context, the term "error" as "something produced by mistake." The term "defect" is defined by the Merriam Webster's dictionary as "an imperfection that impairs the worth or utility." The term "fault" is defined by the Merriam Webster's dictionary as "a physical or intellectual imperfection or impairment." IEEE defines "fault" as "an incorrect step, process, or data definition in a computer program." Notice that this definition is close to the definition of the term "error." The term "failure" is defined by IEEE as "the inability of a system or component to perform its required functions within specified performance requirements." Merriam Webster's dictionary defines the term "failure" as "an omission of occurrence or performance."

Now, allow me to make it more lucid for you. An error is a mistake committed by a programmer in the program. A defect is not a commission of a mistake but an omission of something that could have been done. A fault is something that occurred during the execution of a program due to an error or a defect in the program that led to abruptly stopping the execution of the program. A failure is the abrupt stopping of the program execution. A failure can cause unforeseen problems to the data.

Errors and defects are the root cause of faults and failures. In the earlier days of batch programming, the errors were expected only in our programs. The OS just executed the program. But now with multi-tiered architectures, the faults could develop due to a defect in the program, third-party code libraries, any of the application tiers and incompatibility between the tiers, or the utilities loaded on the server or the client machines. While it is beyond us to check the third-party code libraries and the incompatibilities, we ought to take great care during our code development so that no errors or defects remain in our programs. In fact, third-party code libraries are also sets of programs written by programmers!

Now let us discuss how to prevent errors and defects from lurking in our programs. Anybody can write a computer program, but it takes a good programmer to develop code that is error and defect free.

Facilities Provided by the OS to Handle Faults

When a computer encounters a fault during a program execution, the following sequence of events take place:

1. A fault is developed when the OS encounters a program instruction that produces an impossible result or is impossible to execute and for which the OS has no facility to handle the event. For example, if the denominator in a division operation becomes zero, it produces an impossible result, resulting in an error. The OS has no instruction in its math-processing module on how to handle a division-by-zero event.
2. When a fault is encountered, the OS refers to its error-handling subsystem.
3. The error-handling subsystem of the OS initiates its instructions that initiate the following steps:
   a. The first aspect is to locate the number from its list of errors and pass it on to the program under execution and awaits a response on how to handle the situation.
   b. If the program has a routine for handling the error, it decodes the error and executes the instructions provided in the program to handle the error.
   c. If the program has no error-handling routine, then the program cannot return a meaningful response.
   d. The OS closes the execution of the program and releases the resources held by the program.
   e. The user gets a message from the OS that corresponds to the error number generated earlier.

We programmers should not allow this situation. We ought to invariably include an error-handling routine in every program we develop.

The OS expects faults to develop and result in failures of the programs in execution. The failure of the program should not be allowed to affect the functioning of the computer and hinder the performance of other programs in execution. So, whenever a program fails, that is, the program encounters a fault from which it cannot recover, the OS performs the following actions:

1. It moves the program to wait state.
2. The OS looks at the next instruction of the failed program.
   a. If the next instruction is an error-handling instruction, it will fetch the instruction and place the error information (error number and error message) in the program data area for use by the error-handling instruction. In the next iteration, this instruction will be executed.
   b. If the next instruction is not an error-handling instruction, the OS sends the error message to the user, closes the program, terminates the session, terminates any open connections to databases, and releases the resources held by the failed program.
3. The OS removes the program from the list of programs in execution. The removal is abrupt and does not handle the pending transactions.

If a program is aborted (abruptly closed), depending on the OS and the data storage we used, the following consequences can happen:

1. If there are any flat files that are open, they could be damaged or some data may be lost.
2. If any transaction is being processed, it may be lost or partially saved, resulting in issues of data integrity.
3. If the backend DBMS is a weak one, the table may get corrupted.
4. The user may have already paid the money, in scenarios such as ticket booking, but could not get the desired result. This can lead to a feud between the user and the organization. It can even lead to legal hassles.
5. If a parameter file is impacted, we may have to rebuild it. This can result in our application being down for some time and result in loss of revenue.

Therefore, we need to ensure that every error is handled appropriately by our program.

Errors and Defects

Compilers just ensure that the program adheres to the rules of the programming language and translates the source code statements to executable code. They check for correctness against syntax rules but not against logical errors. Here are the sources of errors in addition to errors in our source code:

1. While the OS is thoroughly tested both inside the development facility and in the field by prospective users in pre-release beta testing, still a few errors lurk inside the OS. These errors lurk inside that portion of the OS that is not used by the general users. The advanced users are few, and the advanced features of the OS are not frequently used. Therefore, when you use advanced features, the OS is likely to throw up errors.
2. It has become commonplace to use third-party code libraries for specialist functionality like file management, database connectivity, rules processing, and so on. These code libraries could have some bugs that could throw up errors when our code tries to utilize that faulty code.
3. The framework or the app server are also built on large code that may have a few lurking bugs. When our code tries to use that portion of faulty code, errors can throw up.
4. The DBMS packages we use for storing and retrieving data also could have a few undetected errors that can throw up errors.
5. We have networking hardware and software that can generate faults and cause errors in our program execution.
6. It is widely accepted that the best quality that can be humanly achieved is of a 6-sigma level in which only 3 bugs can still lurk in a million opportunities; that is, in a million lines of source code, there will still be 3 undetected bugs!

Since it is not possible for us to avoid those errors, we simply have to write a routine to catch those errors and then allow the user an opportunity to smoothly exit that functionality and do something else. We need to strive to avoid program aborts (abrupt closures of the programs) to protect the integrity of our application data and programs.

Errors are usually classified into three levels:

1. *Minor errors*: Minor errors are those errors that do not cause obstruction to processing or functionality but can cause a nuisance to the users while using the application. They can be wrong spellings, wrong placement of labels, insufficient space for entering data, and such other aspects. They certainly cause inconvenience and irritate the user but have no negative effect on the functionality. While the users can continue to use the application, we need to improve the program by removing the errors and rerelease the application once again.

2. *Major errors*: Major errors are those that impact the results of processing or cause obstructions to using the application. Inaccurate results, or diminished precision of results, cause a fault that is recoverable (for example, the result is truncated) and such other errors. The OS allows us to recover from the error. The OS will not abort the program. The users can continue to use the application by isolating the erroneous functionality. This isolation has to be achieved by manual means, which is not desirable. We need to immediately issue a patch if possible, to correct the errors or fix the bugs and rerelease the application as soon as possible.

3. *Critical errors*: Critical errors are those that cause the program failure. The OS cannot proceed further in the execution of the program. The OS aborts the program if the program does not provide for error-handling instructions. When critical errors surface in the program, we need to remove it and fix the bugs immediately. We should not allow the users to continue using the application. We need to rerelease the application only after the bugs are fixed and tested.

Error handling has two sections, namely, error prevention and error detection.

Error Prevention

To prevent errors from causing faults, we need to foresee all opportunities in our programs that can cause a fault and take preventive actions. Here are some areas when errors can raise faults:

1. Arithmetic operations:
   a. The example I have been repeating is the denominator in a division operator becoming zero. Therefore, in our programs, it is essential to split the mathematical formula into two or three parts in our program. We need to write a statement for the portion that comes before the division operation; then the division operation; then the remaining portion of the formula. We need to ensure that each division operation must be in a separate statement. We also need to write a statement before the division statement to check the value of the denominator and send an error message to the user if it becomes zero. It is

better to have a common routine for this purpose and call it in our programs whenever we encounter a division operation.

b. *Negative numbers*: Arithmetic operations may sometimes result in negative numbers. Negative numbers can cause faults. For example, if we are finding a square root of a number that is a derived value from an arithmetic operation in the previous statement, it can result in a fault and lead to failure, aborting the program execution. Therefore, it is essential to check for the sign of the number before subjecting it to square root, cube root, or any other such operation to ensure that a fault is prevented.

c. *Multiplication*: When we multiply two positive numbers, the resulting value can become too large for the receiving variable to accommodate. In most development platforms, the receiving variable truncates the value if it is too large to accommodate. This leads to an erroneous result. While multiplication does not result in a fault and failure, it has the potential to produce an erroneous result.

d. *Size of receiving variable*: The result of the arithmetical operation is assigned to some variable. We need to ensure that this receiving variable is large enough to receive the result.

2. *Loops—interminable*: Finite loops based on a counter usually do not cause any issues, but condition-dependent loops can sometimes become infinite loops and hang up the system. An advanced OS can detect an infinite loop and terminate the program execution, but most others freeze the system and it may need a reboot in microcomputers. To prevent faults, we need to ensure that the loop will certainly terminate. We have to formulate the condition carefully and insert a mechanism inside the loop that makes the condition to become TRUE or FALSE, as the case may be.

3. *Opening files*: We open a file that is expected to be in the location specified or derived by the program. Sometimes, due to some reason, the file may not be there altogether, or its name could have changed for some reason. In such cases, it results in a fault and failure. More often than not, we write programs assuming that the file open operation will not fail. Therefore, to prevent file-opening errors, we need to check if the operation has been successful and if it is not successful, we need to give a suitable message and move the user to some other transaction in the system.

4. *Connecting databases*: In the present day, the database may not be on the same computer on which our program is running. It is usually on a separate machine connected through the Internet. The database connection can fail if the Internet connectivity fails. Internet connectivity can fail if there is a hardware or software problem. The connection can also fail if the database is corrupted, even though such a possibility is very slim. Therefore, we need to check for the success of the connect-database operation before we move on to another program statement to ensure that the database is indeed connected. If there is a failure, we need to flash an appropriate message to the user and navigate the system to another transaction.

5. *Empty tables*: This is another frequent error that we encounter, especially immediately upon installation and during the first use. As soon as we open a database table or a flat file, we begin the statement with the statement "While not EOF" to process the records. Now, if the file or table is empty, it can result in a fault and a failure. Therefore, we need to ensure that there are records, using appropriate statements, in the file or table before issuing the statement "While not EOF" to prevent faults.

6. *Machine readiness*: In programs that interact with machines, like the printer, a CNC machine, or any other machine controlled by the computer, the machine may not be ready for a variety of reasons. When our program interacts with the machine, we must check for errors reported by the machine, flash a suitable message whenever the machine throws up an error, and navigate the user to another functionality.

7. *Deadlock*: A deadlock is a serious error that is not easily detectable especially in a microcomputer OS. This can happen only in the case of dedicated resources. Some resources like the flat files can be opened by only one program at any given instance of time. Earlier printers were also dedicated, but with the development of the SPOOL (Simultaneous Peripherals Operation On Line) facility, this problem was solved. In this, instead of allowing the program to send output directly to the printer, the OS redirects it to a file and prints it whenever the printer is free. The following bullets will explain how a deadlock can occur:

   a. Two (or more) programs are running concurrently.

   b. Program A opens a flat file X.

   c. Program B opens a flat file Y.

   d. Program A tries to open file Y for some reference but has to wait because program B is using it.

   e. Program B tries to open file X for some reference but could not open it because program A is using it.

   f. Now, both programs wait indefinitely!

   g. This situation is called a deadlock.

   h. How do we detect a deadlock? Unless the OS has the facility, it is not possible to be detected by our program.

   i. To solve it programmatically, we need to time out our program. If the required resource is not made available within a specified amount of time, we need to navigate the user to another functionality, giving an appropriate message to try again after some time lapse.

Whenever we try to perform these operations in our program, we need to write appropriate statements to catch the error. These operations may not cause any issue in most cases, but we need to catch the error in those few cases. Therefore, we need to include error-handling statements at all these places likely to throw up errors. In addition to the situations mentioned earlier, there could be situation-specific opportunities for causing faults, and we need to include suitable error-catching statements.

Handling Errors

First of all, we need to write an error-catching statement after a statement that is likely to throw up errors. Some of the possible actions that are likely to throw up errors are described in the previous section. Usually, the error catching statement would begin with "On error," but it could be different in different development platforms. Usually, the error-catching statement directs the program execution to a subprogram that handles the error. Error-handling is the only scenario when the "Goto" statement is permitted in good programming practices! We also pass some parameters to the error-handling subroutine, which includes the error number returned by the OS and other data like the subprogram in which the error was caught and the next subprogram to which the user would be navigated to and so on. Here is an example of an error-handling instruction set:

```
a=b/c
On error goto errorhandler(errno, pgmname)
```

Let us assume a, b, and c are numeric variables. If c becomes zero, then an error will be thrown up. The "on error" statement will pass on control to the subprogram, "errorhandler," which needs the error number returned by the OS and the program name as the parameters. Now the errorhandler subprogram would look something like this:

```
Errorhandler(int numer, string pgname)
{
        switch numer
                case 1
                {
                        msgboxOK ("Dision by zero in subprogram", pgname);
                        goto mainmenupgm;
                case 2
                        ....
                case n
                        ....
                default
                        ...
                }
}
```

In the earlier program, we assumed that the error number returned by the OS was one for division by zero. Then, we display a message box with only the OK button. When the user clicks the OK button, we navigate the user to the main menu form. Of course, the earlier program is not proper syntax. It is shown for illustration purposes only.

We used a switch—case statement. We can put all the error numbers and display different messages for each of the error numbers. It is a better programming practice to retrieve the error messages from a database table rather than hardcoding as shown here.

Possible Alternative Actions for Error Handling

Whenever we encounter an error and we catch it, we can take the following actions from the error-handler subprogram:

1. Close the program and allow the user to restart it. While so doing, we ought to ensure that we navigate the program execution to the program closure subprogram, which will close all open files, database tables, database connections, and all

such actions that are necessary to protect the integrity of data and other artifacts. This is not the best alternative because the user gets frustrated, but sometimes we may not have any other alternative, and only in such circumstances do we need to resort to this action.

2. If it is feasible, we need to flash a message to the user, give the user an opportunity to rectify the situation, and redo the action once again. This is the best alternative, but it is not possible in all cases.

3. We can flash an appropriate message, then allow the user to read it, and once he/she acknowledges the message, navigate the user to the program from where the user came to this current functionality. This may not always be possible.

4. We can flash an appropriate message, then allow the user to read it, and once he/she acknowledges the message, close the current program and navigate the user to the opening page or landing page. In this case, we do not force the user to restart the program, and we allow the user to close the program in a smooth manner to protect the integrity of the data and the program artifacts.

5. In all cases, we need to send an appropriate message to the application administrator so that the fault can be investigated and corrected. In some scenarios, we may need to obtain the permission of the user before sending such a message. It is a better practice to obtain user approval, as the user may know the exact cause of the failure and he/she may take appropriate action without troubling the application administrator.

Handling User Mistakes

Users commit mistakes while using our applications. Most of them use our application the way it should be used most of the time. Very few would misuse it intentionally; most would misuse it occasionally and unintentionally. But the fact is that our application will be misused. The misuse, intentionally or unintentionally, will cause failures and integrity issues. We need to prevent misuse by the users. Here are the likely occasions in which users can misuse the application that we need to prevent.

Clicking the wrong button: This is a frequent occurrence that can cause severe damage unless it is prevented.

1. Clicking the delete button instead of cancel button can cause something to be deleted. To prevent any damage to data, we need to display a confirmatory message asking for the confirmation of the delete action, and only when the user confirms the delete action should we perform the delete action as designed.

2. Clicking the discard button instead of save button. In this case, too, we need to display a confirmatory message and perform the designed action only when the user confirms the action.

3. Clicking the save button instead of the discard button. Usually we do not take confirmation from the user for save action, as the saved information can

always be deleted, but in special circumstances, we need to take confirmation. For example, in a shopping application or ticket-booking application, it is very important to take confirmation before we book the order or the ticket. In such conditions when transaction of money is involved, we need to display a confirmatory message and obtain user confirmation before we proceed with the save action.

4. Clicking on the save or delete button without entering the key information. When entering data, we provide a save button to save the data into the database table. Sometimes the users click the save button without entering any data! Similarly, before deleting information, based on a key value, we display the information that is proposed to be deleted. But sometimes, the users click the delete button even before the information is retrieved. We handle this type of error in two ways:

    a. The first mechanism is to disable the button until the information is present on the screen. In the case of save button, we ensure that all mandatory data is entered before we enable the save button. In the case of delete action, we ensure that the key information is displayed on the screen, at a minimum, before we enable the delete button.

    b. The second option is to carry out a check for the adequacy of the confirmation on the screen before proceeding with the specified action.

    c. In any case, we do not proceed to the action without checking for the information being present on the screen.

Entering the wrong data in the controls is a major misuse by the users: This is a common problem in data-entry screens. Users enter alphanumeric data in numeric fields and numeric data in alphabetic fields. Most development platforms do not automatically prevent this occurrence. Some provide a facility to specify the format for data entry and prevent the user from entering wrong data; others do not. Here are some places where this happens frequently:

1. Entering alphanumeric characters in numeric fields is a very frequent occurrence. We need to develop a subprogram to check numeric fields for disallowed characters and call it whenever data is entered into numeric fields. We need to call this routine during the keypress event so that disallowed characters are not permitted into the control.

2. *Two decimal points or more in a numeric field*: This happens occasionally, and we need to make this a part of the numeric field-checking routine and call it during the keypress event.

3. *Date*: Dates are a special type of numeric fields. There are logical relationships between the date, month, and year fields. Of course, we can use a date picker control and prevent any misuse. Sometimes, we use three separate numeric fields, in which case, we need to check each field for logically correct data by checking the relationship between the month and the date as well as for leap year, in case of a date in the month of February.

4. *Wrong characters in alphabetic fields*: In fields capturing the names of human beings, the only characters permitted other than alphabets are a blank space between parts of a name and a period character as suffix to words like Jr. and Sr. Numeric and other characters are not permitted. We need to write a

subprogram for checking such fields and call it during the keypress event to prevent wrong data being entered into the controls and as a sequel into our database.

Clicking on the empty areas: What happens if we click on the empty area on a form or a frame of a grid? Most development platforms do nothing. They take no action. But I have seen that it initiates some action in a few cases! Frames, forms, and such other controls provide for programming the click or double-click actions for us to use. If we leave that alone, no action is performed, but if we put in some statements in such places, it may cause problems during the runtime. My suggestion is not to program such events unless it has a specific purpose and is so designed.

Canceling the action midway: On many occasions, the users leave the transaction unfinished or cancel the action midway. While we cannot prevent the users from so doing, we need to put in actions in our programs so that such actions do not damage our data integrity. For leaving the transaction unfinished, we need to time out the session and not save it. For the cancelation, we just need to take a confirmation before we actually cancel the transaction. But in both the cases, we should not save the data to our database.

Responsibility for Error Handling

For developing programs, two agencies are responsible: one is the designer and the other is the programmer. It has been a debate for some time as to where the border lies between the programmer and the designer. There are different schools of thought on this aspect. Some would say that the designer is entirely responsible, while some say that the role of the designer is superfluous. The programmer is entirely responsible for the program. In my humble opinion, the designer is responsible for designing the program, and the programmer is responsible for implementing it in the most efficient way. Should we design every small detail? I would say not. The organization has to step in and develop programming standards, which include error-trapping guidelines for all foreseeable errors and user mistakes. For each scenario and domain, there would be special requirements for error-trapping and these have to be designed. Again, there are some special occasions when the programmer would not be able to perceive the error possibility. For example, deadlock is one thing that is beyond the programmer. A designer has to perceive the possibility and design necessary constructs to prevent or detect deadlocks. Similarly, concurrency control is another scenario that a programmer will not be able to foresee. Just a brief explanation of the concurrency control problem:

1. There is one ticket available in a ticket booking application.
2. Two users bid for the same ticket at almost the same time, and the system shows it as available to both.
3. The application initiates the money collection action from both the users.
4. The user whose payment is collected first gets the ticket.
5. The second user is denied the ticket, but the money is taken! His money needs to be returned.

6. In some cases, the ticket is allotted first and then the money collection is initiated. In some rare cases, the payment collection may fail from the user to whom the ticket is allotted. Here we are left with an unsold ticket and a dissatisfied customer in the user who did not get the ticket!

Then there is one more scenario. In report generation, on-the-fly generation from the database has become the norm. In some cases, the report generation may take a significantly long time and the transactions may keep taking place. Before the report is completed, some new records may be inserted and some records may be deleted. By the time the report is produced, the data might have changed and the report may have become erroneous.

These scenarios cannot be visualized by the programmer. These cannot be made part of the coding standards. Of course, they need to be made a part of the design guidelines. These error conditions have to be designed before a programmer can implement them.

Thus, the onus for handling the errors rests on the shoulders of

1. The organization, to develop standards for coding as well as for designing.
2. The designers, to foresee and visualize all possible errors and design them.
3. The programmers, to implement all standards and guidelines for preventing errors and then trapping the errors that can arise during the program execution.

13

Inter-Program Communication

Introduction

Inter-program communication (IPC) is communication between two or more programs in execution. It is also referred to as inter-process communication, as a program under execution is called a process for the CPU. For example, when we print a document on a printer, our word processing program is communicating with the device driver program that drives the printer. This is much more of a requirement when we develop programs for system software like the operating system. Communication can happen in many ways between programs. One way is to pass parameters while calling another program, handing over the execution, and receiving the results. But if two programs have to continuously exchange information between them, we need to use other techniques. This chapter discusses these techniques.

When a program is loaded into the RAM for execution, the OS makes that RAM inaccessible to other programs. This is to protect the integrity of the program space and to prevent other processes from modifying the contents of the RAM without authorization. If other programs can access and modify the data of another program, then hackers will have a field day and computers will become a joke in terms of the accuracy of results. But the C family programming languages provide a facility to access any location in RAM using pointers. Using pointers, we can access any location in the RAM by giving the address of the location of the RAM. That is, pointers can access any memory location, provided it is not allocated to any other program! If a program tries to access a memory location that is already allocated to another program, then an unrecoverable error results, aborting the program. But in multi-user OS computers, even that will be prevented if that location is not allocated to the program trying to access a specific location using pointers. Simply stated, the OS builds fortifications around a program's space in the RAM. The memory space of a program is impregnable and cannot be accessed by any other program running on the computer or even from other computers on the network.

But in building applications, especially in the real-time programming and machine-control systems, programs occasionally need to communicate with each other. Usually we pass some data items from one program to the other program as parameters. We have well-established means for communicating between programs if they are running on different computers using networking, but if the communication is needed between two programs running on the same computer concurrently, we need different means. We can achieve this easily by placing a file or a database table on the disk that can be accessed by both the programs. But the time taken to open a file or a database table is much longer and cannot be relied upon in the real-time software applications. Therefore, the inter-program communication needs to take place via the RAM. Recognizing this need, multi-user operating systems provide for a facility generally referred to as shared memory. Different operating systems provide different kinds of shared memory. The shared memory needs to be declared as a variable or an array of variables. The C programming language provides a specific keyword to declare a variable as shared memory. Here are the general rules to be adhered to while using shared memory.

Inter-Program Communication through Disk Files

Inter-program communication through disk files or tables is the easiest and the simplest form of communication. The program that needs to communicate data to another program stores the data in a flat file or a database table. The program needing information simply opens that file or table and reads. There are some disadvantages in this simple technique. The program that needs to send information would not know that another program is looking for information until it receives an interrupt. The receiving program would not know that the needed information is written in the file or table unless an interrupt is placed on it. This is workable if both the programs are not in execution simultaneously. Still, we can achieve inter-program communication using this technique by continuously checking the files or tables at intervals. But, it would waste a lot of precious execution time of the CPU in polling for the information. It also takes more time in accessing the files or tables. Compared to access, read and write times in disk files or tables is much slower compared to those of the RAM. Hence, we rarely use this technique for inter-program communication.

Facilities Available for Inter-Program Communication through RAM

IPC is achieved using shared memory. Shared memory is a specific amount of RAM that is dedicated for sharing and transferring data/information between processes. This RAM is accessible by any process as necessary, just as a file or database table is accessed by any program. It is implemented in many ways. Here is the way the shared memory works:

1. The shared memory has to be declared before it can be used. In some cases, the program needs to declare it, and, in some cases, the system administrator needs to set up the shared memory.
2. Once declared and set up, the shared memory becomes accessible by other processes.
3. The program needing to access the shared memory needs to connect to it like opening a file. Each OS provides its own set of statements for connecting to shared memory. The connection could be to read or to write or both.
4. Once connected, the program can read from the shared memory as well as write into it depending the type of connection.

5. Once the action is completed, the program needs to disconnect from the shared memory. Of course, some operating systems allow multiple programs to stay connected with the shared memory but not all.

6. In most cases, however, only one program can access the shared memory. That is, even if multiple programs are connected to the shared memory, only one can read or write the shared memory at any given point in time.

We have the following shared-memory facilities for IPC:

Message queue: A message queue is a specific amount of RAM set aside for providing messages between processes. A message contains multiple data items. A queue can accommodate multiple messages, and each message has a specific amount of space to accommodate the message. When a process needing data from the message queue connects, the message at the head of the queue will be delivered to it, then all other messages are moved forward by one slot. The messages are preserved in the queue until it is read. Any process can put its message in the queue, and any process can read that message. These queues are implemented by the programming language, and we need to study the concerned manual to understand the facilities offered by the message queue before we can use it in our programs. Here are the typical features of message queues:

1. To enable message passing, the OS needs to have a built-in message queue facility that handles the message being passed on by the processes (programs in execution). We need to learn about this facility by studying the advanced features of the OS manual.

2. One of the programs utilizing the message queue needs to create and manage the message queue, issuing appropriate statements. Then the message queue becomes available for other programs.

3. Each of the subordinate programs needs to declare the message queue in their programs before using it.

4. Each message has to have a specific length, which depends on the OS that sets the maximum length and the declaration in the program that sends the message.

5. Each message needs to include the process ID (PID) at the beginning of the message. The OS will use this ID to direct the message to the right process.

6. Before using the message queue, the sending program needs to initialize the message queue programmatically.

7. Using proper program statements, the program can place the message on the message queue.

8. The OS will then place an interrupt on the process that is the destination for the message in the message queue.

9. The destination process initiates the process for receiving the message from the message queue.

10. The transfer can happen in synchronous mode (the receiving process halts its execution until the transfer is completed) or in asynchronous mode (the receiving process can continue its execution while the message transfer is going on).

11. If the transfer could not be completed for any reason, then the OS places the interrupt again and resends the message to the destination process once again. The OS keeps doing so until it receives a successful transfer signal from the destination process.

12. Once the transfer is completed, the receiving process sends a signal indicating that the transfer completed successfully. Then the OS will remove the message from the message queue.

As programmers, we need to include statements for composing the message concatenating the data items together along with the destination process ID, initializing the message queue, then issuing the statement for sending the message to the message queue. The rest is handled by the OS. I have described briefly about the message queue facility. You need to study it in depth before attempting to utilize message queues in your application.

Message passing: Message passing is a technique in object-oriented applications to invoke another program to take some action. In other computers, we need to make a system call to invoke another program. Message passing is akin to calling a subroutine and passing parameters. In this case, the main program halts until the subroutine executes and sends the results back. In message passing, the sending program sends some data to the receiving program and both can continue to execute concurrently. Message passing allows programmers to call another object/program without losing control to the called object/program, and once the invoked object/program completes execution, the results, if any, are received through another message. The message can contain multiple data items. The sending program and the receiving program can execute concurrently. The message is to be formed by the sending program and, using the appropriate keywords and syntax, it then sends the message to the receiving program. This sending can happen either in synchronous or asynchronous mode. We need to learn the specific programming language before writing programs using message passing facility. This facility is available in C++ as well as Java.

Semaphore: Semaphores are of two types, namely counting semaphores and toggling semaphores. A counting semaphore is an integer and a toggling semaphore is a bit that can be either zero or one.

1. *Counting semaphore*: These kind of semaphores are used to show the number of units of a resource are available. For example, printers that are connected to the computer are recorded here. Initially, the OS sets this semaphore to show all the printers that are connected to the computer. When a printer is under use by a process, then this semaphore would be decremented, and if a process releases the printer, then this semaphore would be incremented. This semaphore shows only the number of resources that are free, but it will not show which process is using which resource. We need to use some other mechanism to do that.

2. A toggling semaphore is used to show if a resource is available or is locked with another process. For example, if the printer is under use, then this semaphore is toggled to 1 indicating that the printer is not available. When the printer becomes free, it will be toggled to zero to show that the printer is available. The programs read this semaphore and take action as programmed.

The following are the operations performed on semaphores:

1. One of the programs intending to use semaphores needs to declare and initialize the semaphore. Usually, it is the program that controls a device or a resource. Then all other programs can use it. Initialization includes the setting the initial value as well as the permissions to read, write, or both.

2. The subordinate programs can read the semaphore.

3. In counting semaphores, the subordinate programs can either increment or decrement the semaphore.

4. In toggling semaphores, the subordinate programs can toggle the semaphore.

We need to study the programming language manual to properly understand and learn the keywords available in that language before we attempt to implement semaphores in our programs.

Shared memory: Shared memory is like a data file that is resident in the RAM. This facility, if allowed by the OS, needs to be used programmatically. Here are the steps in using shared memory:

1. One of the programs needs to declare and set up the shared memory. The setting-up program specifies the amount of RAM and the type of variables that can be stored in that RAM, including their format.

2. All the other programs that intend to use the shared memory need to declare it exactly in the same way as the setting program defined it.

    a. The size and data type of the shared memory must be same in all programs.

    b. The name used for the shared memory needs to be the same in all programs.

3. The program that intends to use the declared shared memory needs to connect to it using the statements provided by the programming language. The connection needs to specify if the connection is for read, write, or both.

4. Shared memory is like a flat file, that is, only one process can access it at any given time. Multiple processes cannot access it at the same time.

5. Shared memory is implemented in different ways by the different OS. Therefore, we need to study the programming manual and the OS manual to understand, learn, and write programs using the shared memory.

6. The operations associated with shared memory are as follows:

    a. The program needs to issue a statement attaching the program to the declared shared memory.

    b. Open (or enter) the shared memory.

    c. Read, write, or modify the contents of the shared memory as desired.

    d. Close (or exit) the shared memory.

    e. When the operations needed of the shared memory are completed, or before the closing the program, we need to detach the program from the shared memory.

One precaution we need to take is to open the shared memory just before acting upon it and close it immediately thereafter. When a program is manipulating shared memory, other programs cannot manipulate it. Shared memory is like a shared toilet. At any given moment of time, only one entity can use it, but multiple entities can use it one after the other.

14

Coding, Debugging, and Performance Tuning

Introduction

Having learnt the basics of computer programming, including various statements that get the computer to process the data the way it is designed, let us now discuss the nuts and bolts of getting them all together to code the program, make it work, remove the lurking bugs, and then fine tune its performance for release to the users for productive use. In this chapter, we will discuss these aspects.

Coding

In computer parlance, "code," in its verb form, and "coding" are the words we use to denote the work of actually writing the program, putting together all the statements necessary to achieve the functionality assigned to the program. The word "code" in its noun form is used to denote the program statements in a computer program. Basically, we are codifying the algorithm for the computer to decipher it and process the data. The word "code," or "coding," is selected because we are basically putting the algorithm in a code that is understandable to the computer so it can process the instructions. If a layperson untrained in computer programming reads the program, it simply looks like gibberish written in English.

It also includes removing bugs as and when necessary. In Chapter 4, we discussed the basics of computer programs, and in Chapter 20, we will be discussing the programming standards. Right now, let us discuss the programs and programming.

As noted in Chapter 4, a computer program is a series of instructions to the computer that tell the computer what to do. A computer is a diligent machine but not an intelligent one. It cannot determine if an instruction is out of order. It just processes the instructions in the order they are provided in the program. Therefore, it behooves on us to ensure that the instructions are given in the right sequence. Here is how we code the program. We follow a structured way to code the programs. It is generally referred to as the "program structure."

The program structure defines how a computer program needs to be coded. A program would have the following structure, normally:

1. *Form load*: In event-oriented GUI programming, all the processing operations are tied to a screen. Therefore, all programs are connected to the events on the screen. As soon as the user selects the option of running our software by clicking the

option from a menu, we need to load an initial screen. This is usually referred to as the "landing page/screen" or the "opening page/screen." We load this screen first. This will be the first routine to be programmed or rather, it will be the screen that will be invoked when a user clicks the option in the menu that triggers execution of our software. All the other screens are invoked from this screen by clicking the appropriate option or a control. If this screen has some controls on it, then each control would have many events that need to be programmed.

2. *Header routines*: In the C language family including Java, header statements are allowed. These can be declarations or defining formulas as functions and so on. We need to write them first whenever we include header routines in a program.

3. *Program beginning*: Every programming language has a specific beginning that is denoted simply by a single word such as "Main()" or "Identification Division" or something like that. We have to write what is expected of it. In most cases, we do not need to write anything with this statement. We just write the required statement to tell the computer that this is a main program (not a subroutine). This statement needs to be on a separate line in a standalone manner.

4. *Program header*: All professionally managed software development organizations mandate that a program header is written immediately after the statement that begins the program. A program header is a number of commenting statements in which the history of the program is maintained from the first time it was coded and all the maintenance tasks performed on it. It helps to trace any security breach if it takes place. What should be written in the program header is given in Chapter 20 on programming standards. In a GUI environment, we place this header in the event that loads the screen.

5. *Initial routines*: In some programming languages like the RPG, certain subroutines need to be executed before beginning the program code. These statements need to be included immediately after the program header. If there are any predefined routines that need to be performed at the beginning of the program execution, they need to be written immediately after the program header.

6. *Data declarations*: We need to declare all the data items we propose to use inside the program. Many languages allow for the definition of data anywhere inside the program, and some programming languages mandate that all data must be defined at one place. It is better to group all data declarations at one place, as it would make it easy during maintenance and debugging. It is a good programming practice to declare all the data at the top of the program, as it would be helpful to locate any data declaration during debugging or performance-tuning of the program. One question is universal and that is, should I declare one variable per statement or club multiple declarations, albeit of the same data-type, per statement? It is generally better to declare one data item per statement as it helps in maintenance. In case we need to delete or change the type of a data item, it is easier if there is only one data item per statement. Of course, it increases the number of lines but the increase due to declarations does not add to program complexity.

7. *Initialization*: Initialization refers to assigning an initial value to a variable or data item. When a variable is declared in a program, it would be assigned a NULL value. NULL is a separate constant that is neither a space nor a number. When we perform operations, especially arithmetic operations, NULL causes irretrievable errors and program aborts. Therefore, we need to initialize a variable as soon as it

is declared so that we can prevent this kind of fault. Most programming languages allow this operation to be combined with the data declaration statement. We can also initialize a variable at any point in the program. It would be a better practice to initialize the variables soon after declaration or combine the initialization with the declaration. This way, we would eliminate the possibility of using NULL values in arithmetical operations. When we use loops, either finite loops based on counting or condition-based loops, we need to initialize the counting variable of the one used in the condition. More often than not, we forget this initialization operation, which causes infinite loops or accuracy issues in the computations.

8. *Input operations*: We include these operations in the sequence as and when an input is required. After the declarations and initialization, we need to include the input statements because the program needs data for processing. The input may be received from the keyboard, a file, a table, or a port. Here is the place where the processing begins. The first step is to get the inputs, then process them, and in the process, we may need further inputs.

9. *Output operations*: We include these operations as and when data needs to be output from the computer in the chronological sequence appropriate to the action. The outputs may be processed data reports, error messages, confirmatory messages, a beeping sound, or anything that we output to the user. We place these statements at appropriate places in the program depending on the logic followed by the program.

10. *Computational operations*: We include these statements to solve mathematical equations in the chronological sequence they are required. These are written at appropriate places depending on the logic of processing as defined in a design document or flow chart.

11. *Decision-making operations*: Decision-making statements cause the execution to branch away from the next statement based on the outcome of the decision. We need to be careful to ensure that the branching is accurately defined and that the execution thereafter goes to the next chronologically appropriate statement. They are included at appropriate places in the program depending on the need for decisions.

12. *Error-handling routines*: When our program runs, it is likely to throw up some errors. We need to include error trapping and handling routines in our programs. Chapter 20 on programming standards covers the topic of defect prevention in greater detail.

13. *Program ending*: Every program would have an identifiable beginning and an identifiable ending. While the statements at the beginning of the program create the environment for program execution, the statements at the end of the program carry out housekeeping activities so that the next program can run efficiently. We need to close the program systematically. We need to:
    a. Close all open files, tables, and databases that have been opened previously in our program.
    b. Terminate all connections to databases.
    c. Close any input/output devices previously connected to our program.
    d. The programming language may mandate certain statements in the end, and if so, we need to include them.

    e. Any other required statements conforming to the logic of the program like indicating to the user that the program is completed successfully.

    f. In the GUI:

       i. These statements need to be put in the form unload event (or remove the main form).

       ii. Close all open forms that are open.

       iii. Close the login and save the session detail in the relevant file or database table.

In the earlier list, bullets #1 to #7 are to be coded sequentially at the beginning of the program and bullet #13 shall be the last set of statements in the program. The remaining actions need to be programmed as necessary in the body of the program.

In this manner, we write the program statement by statement. In the GUI, we program all the relevant control events as needed by the program logic and complete the program. Each programming language defines its own program structure and we need to abide by it. Once we complete the program, we need to compile it. Compiling will convert the source code (statements written in the selected programming language) into object code (instructions for the computer in the binary language) that can be executed. This compilation process points out syntax errors present in the program. In the bygone days, the compiler used to deliver a list of syntax errors that we needed to check one by one, correct all the errors and resubmit the program for compilation. We needed to do this until all syntax errors were removed. But now in IDEs, the cursor highlights the first erroneous statement, allowing you to correct it. Once we correct it, it takes us to the next error and so on until we correct all the syntax errors.

Once the compilation is completed and the object file is produced, we needed to link it to the libraries used inside the program. In earlier days, we had to do this manually, giving commands and listing the libraries to be linked. But now in IDEs, this is carried out automatically once we select the option to make the EXE file. Once we have linked the libraries and produced the executable file, we need to test it to see if there are any logical or computational errors and then rectify those errors. Let us now discuss testing the developed program.

Testing

It is the testing that delivers the proof of realizing the requirements set for the program and the adherence of the program to its design. It also proves how well these two are realized by our program. Since we wrote the program, it is our child. We have every responsibility to ensure that it works, does what is expected of it, and does not do what is not expected of it. There are two types of testing generally referred to a black-box testing and white-box testing.

In black-box testing, we treat the program as a black box and supply the inputs and look at the outputs. If the outputs are as expected, then the program is working well, and if the results are in error, we go back to the program, look for the reasons for the unexpected results, and then retest the program. We do this until the program delivers the expected results. In giving inputs, we test with a range of inputs, both valid and invalid, to ensure that the program delivers accurate results with the right inputs and blocks the wrong inputs. This type of testing has the disadvantage that we would not be able to give all possible combinations of data inputs that the users in the field would subject our program to. So, we need to do more testing before we certify our program as good. Here are the steps in black-box testing:

1. Prepare test environment:
   a. Create database, if not created already.
   b. Create the tables as necessary, if not created already.
   c. Enter data in the tables as necessary. We need to enter master tables for reference and other transactions tables as necessary.
   d. Create the required flat files and enter the test data using a text editor.
   e. If the program needs to be called from another program, then we need to run that program.
2. Prepare test data:
   a. For each numeric variable to be tested, we need to supply five values in iterations.
      i. One value below the acceptable range—this needs to be rejected by the program.
      ii. One value above the acceptable range—this needs to be rejected by the program.
      iii. One value within the acceptable range—this needs to be accepted by the program.
      iv. One value at the top border of the range or in other words, the highest value that needs to be accepted by the program—this needs to be accepted by the program.
      v. One value at the lowest border of the range or in other words, the lowest value to be accepted by the program—this needs to be accepted by the program.
   b. For each alphanumeric variable, we need to give one value that needs be accepted by the program and one value that has to be rejected by the program.

Once the test environment as well as the test data is ready, we need to run the program and note the actual results and compare them with the expected results. Wherever there is a variance, we need to note the result. Once the program is completely run, we need to go back to the program to locate the errors and correct them, then test it again. We need to iterate the activities of testing and rectification until we can detect no further errors in our program. Then we pass it on to the organizational quality-control department for their testing.

White-box testing subjects our program to thorough testing, or in other words, we test every line of code. White-box testing cannot be carried out the way we carry out black-box testing. We need a debugger or an IDE for carrying out white-box testing. The IDE or debugger would have the following facilities:

1. *Step through the program statement by statement*: Each time we press a specific function key, one statement of the program will be executed. Of course, we may decide to run the program execution unhindered at anytime during the stepping through statements.

2. *Execute the program unhindered*: We run the program as we do in black-box testing. We usually do this once we are confident that there are no errors in the program, just to confirm our confidence, but we can place breakpoints where we suspect an error. This way, we do not need to press the function key multiple times.

3. *Place a breakpoint at any statement*: A breakpoint is marking a statement where the program has to pause execution and allow us to view the contents of variables to determine if there is any inaccuracy. We can place multiple breakpoints in the program as needed.

4. *Pause the program execution*: During the program execution, we can pause the program execution by placing a breakpoint at any statement. When we pause the execution, we can view the contents of variables and see if anything is amiss.

5. *Resume execution*: We can resume the program execution that was paused after we viewed the contents of variables and completed the action for which we paused the program execution. We can either resume the program execution, end the execution, or step through the program, one step at a time.

6. *Change the position of resuming execution*: Once we pause the program execution, we can move the pointer to a previous statement or to a later statement. We can move the next point of resuming execution using the mouse and drag the pointer to a statement from which to resume execution.

7. *End the program execution*: We can also end the program execution at any point of time during the execution of the program. But we need to have paused the execution either using a breakpoint or a completing one step using the stepping key. We usually do this when we have located the issue and wish to correct it.

8. *View the contents of any variable when we pause the execution using a breakpoint or after each step*: When the execution is paused, we can view the contents of variables using the display commands. By viewing the contents of the variables, we can determine if the actual results are as expected.

9. *Change the contents of variable*: If we determine the contents of the variables are at variance with the expected values, we have the option to correct the statement, move back and re-execute the statement, or change the variable value and resume execution forward. Using this facility, we can change the value, move forward, and correct all errors in one go.

10. *Change the program statement*: IDEs usually allow us to change the program statements when we paused the execution so that we can see the results of the changed statement immediately.

We carry out white-box testing using these facilities. This is how to carry out white-box testing:

1. When we begin testing in the IDE, we select the option to move through the program statement by statement. Alternatively, we can place a breakpoint at the very first statement and then step through the program, one statement at a time.

2. Pressing the function key to step statement by statement, we begin from the first statement and then continue until all the statements are executed.

3. After stepping through the first statement, we continue until we come across a control statement like IF...THEN...ELSE. At this point:

a. We allow the execution to take the branch that is allowed by the data and continue stepping through until the last statement of the control statement block is executed.

b. Once all the statements in the control statement block are executed, we move the execution pointer to the statement prior to the first of the control statement block.

c. We change the value of the concerned variable such that the execution takes the other branch of the control statement.

d. We continue stepping through the remaining statements of the control statement block until all the statements in the control statement block in the branch are executed and the last statement of the control statement block is executed.

4. We repeat step 3 for all the possible branches in the control statement block. We need to note that the switch...case control statement can have multiple branches. Similarly, there could be multiple ELSE statements in the IF...ELSE... THEN control statement.

5. When we come across a loop, we need to execute the program in such a way that the execution enters the loop and executes all the statements in the loop statement block. We also need to create a condition that the execution does not enter the loop statements block and execute the program. If we read records from a database table or a flat file, we need to execute in such a manner that we need to execute the loop once with records in the table or file and once with zero records in the file or table.

6. While receiving inputs from the keyboard:

a. We need to execute the program receiving expected inputs to see that the program works as expected with right inputs.

b. We also need to execute the program with wrong inputs to see that the program rejects the wrong inputs as well as retains control of execution without resulting in a fault condition. While giving wrong inputs, we need to give wrong inputs for all the controls one by one and see that the program rejects all the wrong inputs from any of the controls.

c. We also need to execute the program with no inputs at all to see that the program does not move forward until all the mandatory inputs are received.

7. When we receive inputs from other sources like machines and the Internet:

a. We need to execute the program with right inputs.

b. We need to execute the program with wrong inputs.

c. We need to execute the program with the source disconnected.

d. We need to see if the program rejects wrong inputs and alerts the user about the nonexistence of connection with the source.

8. When connecting with the databases, we need to execute the program with a proper connection and with no connection to see that the program handles both situations properly and retains the control of program execution.

9. When delivering outputs as reports:

a. We need to ensure that the results are accurate.

b. We need to ensure that all the records that need to be included are included and that all records that need to be excluded are excluded.

c. We need to manually check that the derived/computed figures, page totals, and grand totals are accurate, comparing the manually arrived figures with those of the figures printed in the report.

d. We need to manually check that the control statistics report matches with the data delivered as output.

10. When delivering outputs to a table or file:

a. We need to see that the destination file or table is indeed populated.

b. All the records expected to be written onto the file or table are indeed written and saved to the file or table.

c. The data stored on the file or table is accurate and no errors have crept in while writing to the file or table.

11. When delivering outputs to a machine, we often do not have access to the machine. We usually conduct this testing using a mock-up or a simulator. We can't expect to test an airplane or a rocket with untested software—right? But once the program is tested on a mock-up and all errors are eliminated, we have to test it on the actual machine one day. Such field testing is usually carried out under rigorous conditions by experienced testers who begin testing only when they are sure that the test will pass. That testing is only confirmatory and no errors are expected. During testing the program for delivering outputs to a machine:

a. We need to ensure that the machine is in such a position and that it causes no damage or injury to anyone in case the program behaves in an unexpected manner during testing.

b. We need to observe the machine behavior to see that the exhibited behavior is as expected.

c. We need to place the machine in error condition such as not being powered up, missing vital supplies such as paper or a tool, and so on to ensure that the program traps the error but retains the control of the program execution without aborting and releasing control to the OS.

d. We also need to disconnect the machine and see if the program is trapping the error and recovering control of program execution.

12. When delivering outputs to the Internet or a local network, we need to test with proper connection to the network as well as with no connection to the network. We need to send the output onto the network but to a known destination over which we have control. If we are sending an email, we need to send to our account. If we are sending a message or making a call to a mobile phone, we need to do it to our phone. We need to see that the sent message is delivered properly and that the call is connected with proper reception. We also need to see that the messages sent on the network or Internet are received properly at the addressed destination.

Software testing is a big topic. A new field of software testing has emerged, and a body of knowledge is being gathered. But even though we are specializing as programmers, the onus of certifying the program as satisfactorily working without any defects rests on our shoulders, and we need to test it. In fact, programmers are the first line of software testers, albeit the fact that they test their own programs only. I have included a brief outline and important aspects of testing here. Interested readers may refer to a good book on software quality assurance.

Debugging

One version of how the word debugging came into being was described in Chapter 4. Here is the more popular version. The word "debugging" is attributed to Rear Admiral Grace Hopper. While she was working on the Howard Aiken-built Mark-II computer (Automatic Sequence Controlled Calculator), she located a moth stuck between the contacts of relay #70 on panel-F on September 9, 1947. Madam Grace Hopper added the caption "First actual case of bug being found." You can see the picture of the log here (Figure 14.1)

I am not sure if it is really written by Madam Grace Hopper, but it is touted so! But of course, the word "bug" to denote a glitch was used by Thomas Alva Edison, too, and much earlier. Here is an excerpt from his papers: *"It has been so in all my inventions. The first step is an intuition and comes with a burst, then difficulties arise—this thing gives out and then that "bugs"—as such little faults and difficulties are called—show themselves and months of intense watching, study, and labor are requisite before commercial success or failure is certainly reached."* It is in a letter from Edison to Puskar, dated November 13, 1878.

Having noted the history behind the word debugging, let us turn to our discussion on removing errors, or "bugs," as we prefer to say, from our programs. We make two kinds of errors in our programs. The first one is syntax errors, which are trapped by the compiler and force us to correct them before running the program. The second category is the logical

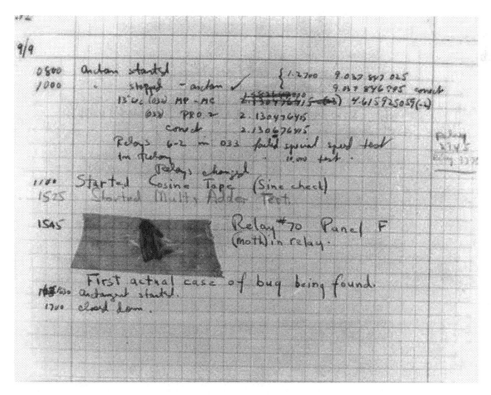

FIGURE 14.1
First-ever bug report.

errors. By logic, I mean that there is some defect in our logic/algorithm. A logical error produces a wrong or inaccurate result. Here are the logical errors we frequently commit:

1. Getting a wrong result in arithmetic operations.
2. Getting wrong data in retrievals.
3. Getting wrong results in printing.

Getting a Wrong Result in Arithmetic Operations

This is a very common error that surfaces in computer programs. We use different symbols for solving algebraic equations manually on paper and for solving them by computer programs. For example, consider this equation:

$$a + b^2 \times c \div d - e$$

In this equation, the arithmetic symbols are between the variables. This is referred to as "infix notation." There are two other notations referred to as "postfix" and "prefix" notations. In infix notation, the arithmetic symbol is located between two variables. In postfix notation, the arithmetic symbol is located immediately after two variables. In prefix notation, the arithmetic symbol precedes two variables. In postfix notation, we write the earlier equation, thus:

$$b^2 c \times d \div a + e -$$

Multiply b^2 by c; then divide the result by d; then add a to the result; and finally subtract e from the result. As you can see, the arithmetic symbol is placed immediately after the two variables on which it is to act. (a b +) is the same as (a + b). Computers convert our arithmetic statements into this form for resolving them.

In prefix notation, the arithmetic symbol is placed before the variables it has to act upon. (a + b) is written as (+ a b).

While solving this equation, we would:

1. Compute the value of b^2.
2. Then multiply the result by c.
3. Then divide the result by d.
4. Then add the result to a.
5. Then subtract e from the result.
6. The result would be our solution.

We do like this because our math teacher taught us BODS (bracket, of, divide, and then subtract or add). That is the precedence order of arithmetic symbols in solving algebraic equations. That is:

1. First, we solve what is inside a parenthesis.
2. Then multiply.

3. Then divide.
4. Then subtract or add.

To make the earlier equation explicitly lucid, we change the equation thus:

$$a + \left(b^2 \times c \div d\right) - e \text{ or } \left(b^2 \times c \div d\right) + a - e$$

Should we simplify it this way? When dealing with human beings, it is not necessary but when you are dealing with a computer, yes, I would say it is a must. The computational errors crop up because we programmers assume that computers are intelligent. Computers by nature are diligent workers and intelligence needs to be programmed in! Who else, apart from us, would do the programming part? Therefore, we need to program that intelligence into the computer. To avoid computational errors, we need to take the following precautions:

1. Use more parenthesis liberally. Never economize on the use of parenthesis.
2. Do not nest more than three sets of parentheses in one equation. It becomes very confusing when debugging the computational errors if you have more than three. If you need to use more than three sets, split the equation into two assignment statements.
3. When programming long equations, split them into multiple assignment statements, than pack the entire equation into one line.
4. Give priority to clarity over brevity while programming arithmetical equations.

Now how do we first catch the computational errors in our programs? We do it like this:

1. We place a breakpoint at the first statement of the assignment statement carrying out computation.
2. From then on, we step through every statement until we reach the final statement arriving at the final result.
3. After executing every statement, we view the result of the computation and check it out manually. If the result is not as expected, then we step back to the erroneous statement, analyze what caused the error, and correct it. We step forward, executing the statement and check the result once again. We do this until the result computed by the program statement is the same as the result we worked out manually.
4. We do this for all computational statements in the program to eliminate all errors.
5. If the computation statements are too long, it will be very tedious for us to check the results. Therefore, it is suggested to split long computational statements into multiple smaller statements. If you divide, you can rule!

What causes the errors in computational statements? Here is a brief list:

1. Wrong arithmetic operators by oversight—a very common mistake.
2. Not using parenthesis and relying on the natural precedence order of solving the equations.

3. Nesting too many parentheses leading to confusion in arriving at the results. The confusion is not for the computer. It is for us.

4. Making the statement too long without splitting it into multiple smaller statements. When we do so, we forget the parenthesis and order of the operators, and even use wrong variables!

Results Getting Truncated

This, again, is a very common problem, but it is in fact a design problem rather than a programming problem. When two values are multiplied, the result is larger than either value. In an arithmetical equation that has all the arithmetic operators, it is not possible to predict the result beforehand. As we are the people that declare the variable that receives the result of an arithmetic statement, we need to be careful while declaring it. We can take the size of the field in the database where the result is ultimately stored as the size of the variable. But, during white-box testing of the program, we need to give the largest expected values and see if the result is getting truncated. So, when we pause the program after the execution of an arithmetic statement, we need to look for the size of the result in addition to the accuracy of the result. If we detect a size error, we need to step back the declaration statements and correct the size of the variable at hand. But, using this ploy, it would be counterproductive if we declare all variables to their maximum size as it consumes larger amounts of RAM.

Getting Wrong Data in Retrievals

We retrieve data on many occasions and this is one of the common functionalities in programming. But except in some very rare circumstances, the retrieval will be based on some conditions enforced by relational and logical expressions. While we do not make many mistakes in relational expressions, we do commit blunders in logical expressions. While a relational expression compares two variables, a logical expression compares two relational expressions. In relational expressions, the most common mistakes occur at the borders, that is, when we use relational operators <= (less than or equal to), >= (greater than or equal to), and the values of both the variables are just equal to each other. We need to place a breakpoint after each relational expression and check the result, especially when the values of both the variables are equal. We also need to change the values of both the variables so they are equal, execute the statement and then check the result. When we execute a logical expression, the computer resolves all relational expressions and then resolves the logical expression from left to right. There is no set precedence order for solving the logical operators. When we use one logical expression (just two relational expressions), it is very easy to predict the result. But when we use more than three logical expressions in a statement, it becomes very difficult to predict the result. I recommend not using more than three logical expressions (that is four to six relational expressions in all) in a single statement. But life is not fair, and we may need to use more than three in a single statement owing to the peculiarity of the situation. So, in order to debug this logical statement:

1. We need to program the statement using our best logic.

2. Place a breakpoint before the statement containing the logical expression.

3. Change the values of the variables in such a way that we can easily work out the result manually.

4. Step to the statement through the logical expression and execute it.

5. Pause the execution by placing a breakpoint on the statement immediately succeeding the logical expression statement.

6. Check the result and if it is not as expected, then step back to the statement with the logical expression, analyze it, and correct it as necessary.

7. Step back to the statement preceding the logical expression, change the values of the variables, step forward, and execute the logical expression.

8. Repeat these steps as necessary until the result of the logical expression is as expected.

In this manner, we need to debug the errors. Once the relational and logical expressions work without defects, the search will result in retrieving right data.

Getting Wrong Results in Printing

This is another common mistake that occurs while printing out a report. One common mistake is the data getting truncated in printing, even though the field has adequate width to accommodate the result. This happens when we allocate inadequate width for a report variable. In some cases, it is possible to wrap the text in the report, especially if the data item is of alphanumeric type. Numeric data too, perhaps, can use wrap-text facility, but it looks odd. In some cases, like printing bank checks, we should not use such option. So, to debug this kind of mistake, we need to use the highest possible values and print the test report. When an error is uncovered, we need to go back to report design and correct the error by providing an adequate size for the data on the report and then retest until we obtain satisfactory results. In the case of bank check printing, we can't use the actual stationery as such stationery can't be wasted and it has the potential to be misused. Therefore, we print the check details on plain paper and place it on a blank check and ensure that everything is alright.

Another common error is overprinting, especially in the case of preprinted stationery. Instead of printing in the blank space, the value is printed on the preprinted matter. Another error is that the printing is in the blank space but either above or below the space provided for it. This could be because the stationery is mounted incorrectly. Sometimes, it could be an error in the coding of the report. We need to print it on preprinted stationery, if possible, and check and correct the mistake. If it is not possible to print the report on preprinted stationery, we need to print the report on plain stationery and place it on preprinted stationery to inspect the locations of results and ensure the printing is programmed correctly. We need to correct it if any errors are uncovered.

In debugging reports for errors, we cannot place breakpoints and test the report, especially if we design the report using a report-generation tool. If we code the report entirely using programming statements, then we can place a breakpoint in the loop that is programmed for printing the report, check line by line, and then debug. Since most of the reports are coded using report-generation tools, we need to produce the reports and check the report manually. We look for these aspects and correct the mistakes uncovered, if any:

1. Appropriateness of the report headings, page headings, column headings, page footers, and section footers.

2. Spelling errors in any of the headings and footers.

3. Alignment of data in the columns with the column headings.

4. Accuracy of all the computed values on the report.

5. Accuracy of the page totals, section totals, running totals, and grand totals.

6. Adequacy of space provided for the data and totals on the report. We specifically look for any text and computed values being truncated or wrapped around to the next line. These need to be corrected.

7. We specifically look for rounding-off errors, especially in various totals on the report. We check this by ensuring the grand totals with the section totals or the page totals, the page totals with manually computed totals, and so on. When the rounding-off errors occurs, the higher-level totals vary from the sum of lower-level totals.

8. In reports using preprinted stationery, we see if the printed values are properly aligned with their captions and rectify the mistakes.

9. We also see if the selected font size is amenable for easy reading for the intended audience.

10. Any other situation-specific requirements.

Then we rectify all the mistakes and check the report once again, repeating the cycle until we can detect no more defects.

Debugging Erroneous Communication

Communication in the context of programming and debugging is to ensure that what is sent from the program is the same as what is received at the intended destination. To debug the communication program, we use a simulator that will act as the intended destination during debugging. We will be using a mock-up of the actual hardware and software during testing. The simulator is usually a software running in the background that receives communication and gives out the responses commensurate with the messages sent by our program. In debugging our communication programs, we look for the following aspects:

1. Appropriateness of protocol, that is, the communication protocol used in our program is being accepted by the destination and we are receiving appropriate responses.

2. The communication is being received by the intended destination and not ending up at undesired destinations.

3. The communication is rejecting undesired destinations, if this aspect is programmed in our program.

4. The message received by the destination is the same as the message sent by our program.

5. In the case of financial systems, the amount sent by the program is the same as the amount received by the destination. We also check if the action requested is the same as the action performed. That is, if the program asked for deducting the money, the destination actually deducted the money.

6. Any other situation specific aspects that are needed to be checked and debugged.

Performance Tuning

Performance tuning refers to the activities by which we try to improve the response times of our programs thereby improving the performance of our programs and the throughput of the computer.

The performance of a software program is measured by the response time delivered by the program. Response time is the amount of time taken from the time the program is initiated to the start of the response being perceived by the user. If you type a URL in the browser, the response time is the amount of time taken from the time you hit the enter button or clicked the appropriate button until the time when the display begins to appear on the screen. If you are generating a report, response time is from the time you clicked the generate-report button to the time the report begins to appear either on the screen or on the printer. In communication programs, the response time is measured from the time we click the button to send the message to the time it begins sending the message. In a nutshell, the response time is measured from the time the user completes the action and entrusts the data to the computer beginning the action. Unless the user can perceive the beginning of the action, it would appear that the computer is not acting. Obviously, the computer needs time to process some amount of data before it can be presented to the user. Sometimes, the data is of low volume and sometimes data is of very high volume. The amount of processing time taken by the program depends on the volume of data. When the volume of data is very high, we need to devise ways and means to speed up the processing.

Even in business applications where the response time is not critical, it is important to keep the response time shorter. In multiple studies conducted on the user perception, the following information surfaced:

1. A significant chunk of users will leave a website if the response time is longer than just 4 seconds!
2. Users with a little more patience leave the website if the response time is longer than 8 seconds!
3. Most users cannot withstand response times greater than 15 seconds.

No web-based application should have a response time greater than 15 seconds. We, as programmers. should do our bit to keep the response times as low as possible, and performance tuning is the way to do it.

Performance of a program and the response time depends on:

1. Hardware, CPU, RAM, secondary memory, and network speed/bandwidth whose capacity to process data is finite and predefined before our program is coded. They allow a certain number of computer instructions to be processed per unit of time. We were measuring this by the MIPS (millions of instructions per second) and later with Megaflops (millions of floating point operations per second). The CPU speed is measured in GHz (gigahertz per second with each hertz being equal to one cycle of the system clock).
2. The OS is a collection of programs that govern and control the computer hardware. Some OS has more programs and some have less. The OS of a PC would have much less number and complexity of programs than the OS of a mainframe

computer. Even in PC, the competing OS would have different sets of programs with differing complexities. So, the OS would also impact the time taken for processing the data.

3. Other programs running on the computer. In these days, multiple programs are running concurrently on the computer, and our program has to share the computer resources with all those programs. The time allocated to our program depends on the number of programs running concurrently on the OS. The response time deteriorates as the number of programs concurrently running on the computer increases.

Even with all these being in place and out of our scope, we still need to ensure that our program is taking the least possible amount of time for processing the data. For this, we need to ensure that our program does not contain statements that are not really required or contributing to processing the data. The following are the wasteful statements that we need to eliminate from our program:

1. Declaration of data items that are not used in the program.
2. Loops that have more statements than necessary.
3. Too many relational and logical expressions in control statements.
4. Nesting of too many control statements.
5. Splitting the arithmetic statements into too many statements.
6. Opening too many databases and tables concurrently.

Of course, it is granted that we do not commit any of these mistakes willfully. But how do we determine if we did commit those mistakes by oversight? Fortunately, the SDK developers recognized these possibilities and generally include some performance-tuning tools. These are:

1. *Profiling*: The profiling tool is a software tool (also referred to as a profiler) that runs the program under its control and analyzes the program while it is in execution. It gives a wide variety of information about the program including the amount of RAM used, the time consumed by a single statement or a block of statements, the amount of time consumed by the subroutine, the number of times a subroutine was called, the number of times a statement was executed, and so on. This information is generally referred to as the execution profile of the program. We can also use the profiler with a software product to analyze the product during execution. In fact, most web-hosting platforms use a profiling tool to collect information about usage and then to improve the performance of the application. It then gives additional information about the number of times a program was called along with the idle times and so on. Using the information collected by the profiler tool, we inspect the places in the program that are taking more time and improve them. We do the following to improve performance and reduce the response times:

   a. Delete unnecessary declaration of variables.
   b. If it is feasible, we try to use the same variable instead of declaring too many variables, especially for temporary purposes like flags and counters.

    c. We reduce the number of files or database tables open concurrently. We open them only as absolutely necessary and close them immediately after they are used.

    d. For many statements, alternate statements are available in most programming languages that take less time but are complex to use. We use such statements where necessary to reduce the response times.

    e. We inspect loops and see if there is a possibility to reduce the number of statements to improve the performance.

    f. We inspect our control statement blocks to see if we can improve the logical expressions and also the number of statements in the block so that we can reduce the response times.

    g. If performance times are critical as in real-time machine-control systems, we remove all statements included for aesthetic appeal. These statements could be for setting background/foreground/font colors and special effects such as flashing, fireworks, and animation.

    h. Sometimes, the algorithm used may not be the most efficient one. There are always alternatives for almost any algorithm. So, if we find a situation that is not improved by other means, we approach an expert mathematician and take a deep look at the algorithm used in the program and see if it needs improvement. Of course, only we resort to this in extreme cases, as changing the algorithm would result in recoding the program—well, almost!

    i. Any other situation-specific improvements as possible.

2. *Tracing*: Tracing tools give a record of all the events along with the times they consumed while the program is in execution. It will be a long list, as each iteration in all loops or all blocks of control statements would be traced and listed. It will be a pretty long listing and wading through it takes considerable amount of time as well as patience, but it gives valuable information for performance improvement, especially after running it through a profiler and implementing improvements uncovered during program profiling. A trace of a complex program running into hundreds of lines is an impractical thing because of the sheer length. Therefore, in most cases, we initiate tracing only at such suspicious points uncovered by the profiler and terminate the trace at the end of the suspicious block of statements. This would be short and gives valuable information for performance improvement. Tracing is also used to find out if the program execution is being transferred to some other undesirable program due to a malicious statement included innocuously. We use the tracing tool when the profiler uncovers a block of statements that are taking longer times and we could not see why.

3. *Manual*: We use peer-review of code for quality-control purposes, but it can also be used to improve performance. When a senior and expert programmer reviews the code of a program, the inefficiencies built into the program are uncovered. The reviewer will submit a report detailing the opportunities for improvement along with defects. We improve the program performance by effecting those opportunities for improvement suggested by the expert programmer.

Performance Tuning in Databases

Data is critical in program performance. After all, programs receive and store data or retrieve and report data in business applications. If the database is poorly designed, then whatever improvements we make in the program would all be a waste of time. Database tuning is a part of the database design and administration, but I will discuss it very briefly here just for the sake of understanding. Here are the critical aspects of database design that are critical to the response times achieved by the software programs:

1. The data types of the data items that are stored in the tables are very important. The data types declared in the database ought to be the same as those that are available in the programming language. For example, most databases have a data type for "money," or "currency," which has only two digits after the decimal point, but most programming languages have only integer and floating-point data types for numbers. It places an overhead on the program execution to convert or reconvert the data type every time it is used in the program!

2. Data redundancy can be controlled in the databases, but it cannot be totally eliminated, especially in relational databases. We have to carefully design the databases minimizing the data redundancy.

3. When we need to query the database based on criteria, placing the query in the program takes a longer amount of time during the program execution. Databases give facilities to place this query on the database itself, limiting the program to supply the parameters. It is better to use this facility and limit the program to supplying the query parameters and receiving the results. This will reduce the response times.

4. Most modern databases build the required indexes automatically, but if some databases allow us the facility to build indexes, we need to design the indexes carefully so that the data retrieval times are kept to a minimum.

5. Databases provide a facility of views or joins (or logical files) which basically take data from two or more tables on the fly and present it to the program or user. It is better to use this facility than to have a physical table to store the data collected from multiple tables. This would reduce the response times.

We need to entrust the activity of the database design to an expert in database technology and subject the design to quality control activities before using it during the development. Once the database design is frozen and we begin development of the software, it is well-nigh impossible to go back and change the database design, as there would be number of programs that would be needing change.

While coding is carried out by all programmers, including new entrants to programming, the activities of debugging and performance tuning are expert activities and are generally carried out by senior programmers. Coding guidelines provide guidelines so that our programs work efficiently, besides delivering the expected results, but we need to tune the performance of the programs when the response times are critical. We need to gain expertise in these activities.

15

Subroutines

Introduction

After a few years of software development, it was recognized that writing long code is extremely difficult to debug and maintain. Therefore, structured programming was advocated, which primarily involved writing a short main program that calls many subprograms. It is still debated as to how long a long program is. COBOL came up with paragraphs to shorten the length of a program and FORTRAN came up with subroutines to achieve structured programming. All the other languages that came later on had some facility or the other to shorten the main program and have multiple subprograms to achieve the total functionality. A program that cannot be executed on its own is a subprogram, subroutine, function, or method. It needs to be called by another program for execution. Each language has its own label to refer to this facility. Subroutine, Subprogram, Function, Method, Procedure, and Object are some of the names that are used for this facility. By whichever name it may be called, it is simply a facility to hive off some of the code of the main program so that the main program is shorter and easier to understand and maintain. I am using the name "subroutine" in this book to represent all the names used in different programming languages. In the current day GUI programming, all are subroutines. Each event of a control is a subroutine. The main programs are embedded in the forms. All forms are connected to a main form or a landing/home page.

Characteristics of a Subroutine

The following are the characteristics of subroutines:

1. Subroutines are usually embedded inside a main program. The subroutines can also be placed in a library of subroutines in some programming languages. The library is linked during the process of compilation and producing the executable file.

2. A subroutine is assigned a name which the main program and other subroutines use to call it and hand over the control of execution to accomplish a specified functionality.

3. A subroutine is used to deliver one specific functionality. Of course, there is no restriction on the number of functions a subroutine can deliver, but it makes sense

to hive off the second functionality to another subroutine, as the main objective of using the subroutine is to keep it short and easier to debug and maintain. Having multiple functionalities makes the subroutine longer and therefore defeats the very purpose for which subroutines are used.

4. A subroutine is a self-contained program that delivers, normally, one functionality. It can receive arguments (data) from the main program and returns the result back to the calling program.

5. A subroutine can call another subroutine. It is not necessary that only the main program has to call a subroutine. One precaution, however, is necessary. The called subroutine should not call the subroutine that called it. For example, A is a subroutine and it called the subroutine B. The subroutine B can call other subroutines as necessary, but it should not call the subroutine A that called it. If we do that, we enter into an infinite loop and the program crashes or the computer freezes!

6. The subroutine begins with a keyword such as "Sub" or some other word to denote that it is a subroutine. This causes the OS to add this into the processes waiting for execution and keep the main program in wait state until the execution of the subroutine is completed.

7. The subroutine ends with a keyword such as "Return" or something like that to denote the completion of the subroutine. This statement tells the OS that the execution of the subroutine is completed and that it can be removed from the list of processes waiting for execution, as well as to bring the main program back into the list of processes waiting for execution.

8. Usually the subroutine hands the control of execution back to the calling program, but in some languages, the facility to end even the calling program is available. This facility is useful especially when the execution encounters an error from which the program cannot recover and closure is the only option.

9. Subroutines are usually embedded in a main program. An alternative practice is to build a library of subroutines that deliver commonly used functionalities such as checking a text box for valid numeric data item. This library may be built at an organizational level or a project level.

Function

Functions were initially used in FOTRAN programs in addition to subroutines. A function in the context of FORTRAN programming was a single-line program statement in which an arithmetic formula is assigned to a numeric variable. The formula is defined in the usual manner using variables and constants. Then this numeric variable was referred to wherever the formula was needed. In this manner, repeated coding of formula was avoided, and in their place, variables were used. During compile time, these variables were substituted with the actual formulas. That way, FORTRAN reduced the tedium of coding for the programmers.

Then the C language used functions in another way. In the C programming language, all programs are functions. The main program is called the "main" function. To be executed independently, the "main" function is essential. All functions are multi-line functions, unlike in FORTRAN. What is more, C language allows each function to be compiled into an object file independently, which is not an executable file. Further, C language allowed all such object files to be combined into a library too! This library could then be linked to the "main" function at the time of preparing the executable file. That was a great advantage, as a function need not be coded again by another programmer. This is better than the subroutines used by FORTRAN.

All the next-generation programming languages of the C language such as C++, C#, and Java followed this methodology.

Methods

Methods are used in the Java family of programming languages. In Java programming languages, a method is more or less the same as a subroutine. Many preprogrammed methods are made available to the programmers, which can be called by the programmer. I am not going into the details of coding Java methods because this book is not about programming in Java.

Objects and Classes of Object-Oriented Methodology

OOM (Object-Oriented Methodology) uses classes and objects. Objects and classes are like subroutines, but have significant differences. An object can have multiple methods with their own data definitions. The calling programs can call any of the methods contained inside an object. An object can be a mini-library! An object is an instance of a class, or in plain words, an object contains a class or classes inside it. An object can have additional methods besides a class. Objects can call any of the methods inside another object. I do not wish to get into the details of object-oriented programming here. It takes many pages and deviates from the main purpose of this book, which is to introduce you to the general programming concepts. Suffice it to say, objects are also akin to subroutines.

Data Used inside a Subroutine

When a program is executed, the OS allocates space in the RAM to house all the data items declared in the program, and this space is held by the program until it is closed. RAM is a precious resource that is always in short supply. When the subroutine is spawned for execution, the OS allocates space in the RAM to hold its data items, which is released immediately after the subroutines is executed and is closed. This is a great advantage as the RAM need not be dedicated for the entire duration of the program execution. Subroutines reduce the burden on the availability of RAM.

Subroutines can always access all the data items in the calling program without any restriction. Therefore, there is no point in declaring a data item in the subroutine if it is already existing in the memory space of the calling program. It will be duplication and wastes precious RAM. But if the data item in the calling program is holding a value and cannot be released for use by the subroutine, we need to declare the data item in the subroutine. Conversely, if a data item is needed only by the subroutine, there is no point in declaring it in the calling program as it will tie up the RAM for longer duration. While declaring data, we need to ensure that the declared data item is used only in the program in which it is declared.

We discussed about the static and dynamic data types. Static data is that data item that can be used in all subroutines and calling programs. We can declare static data types in subroutines also. Such data items survive the closure of the subroutine and will be released only when its calling program is closed and all of its memory space is released. Dynamic data-type items are declared and are local only to the program in which it is declared as well as all the subroutines called by it. Unless there is a pressing need, we ought to use only dynamic data types. We may declare a static data type only as a last resort only when there is no alternative is available.

Argument Passing

There are two methods to pass on data to subroutines. One way is to declare the variable in the calling program so that the subroutine can make use of the variable to carry out the processing task assigned to it. In this method, the value of the variable in the calling program can be modified by the subroutine. The second method is to pass arguments to the subroutine. This keeps the values of the variables in the calling program to retain their original values. The subroutine receives only the values and uses them to carry out the processing task assigned to it. However, we need to declare the variables inside the subroutine to receive those values and store them until the execution of the subroutine is completed.

The values being sent to by the calling program are referred to as arguments in the main program. The values in the subroutine that receive values from the main program are referred to as the parameters. But, I confess that these terms are not universally accepted. Some may call both parameters or arguments, or may call the values being sent as parameters and the receiving values as arguments!

In most cases, the second method of passing arguments is preferred, as the values of the calling program are not changed by the subroutine in an unpredictable manner. Another reason is the possibility of the subroutine being used at more than one location in the main program or by another subroutine. When the subroutine is called multiple times from different locations in the main program or by different subroutines, the values of the variables in the main program become unpredictable and can produce inaccurate results, and debugging and maintenance becomes a nightmare.

When passing arguments, we need to take these precautions:

The number of data items passed on to the subroutine must be the same as the number of arguments declared in the subroutine. For example, take a look at the following call to a subroutine:

```
int basicpay;
float allowance;
float deductions;
float empsalary;
/* here we read the table and obtain the data and call a subroutine
   to compute the salary*/
empsalary = call Sub_compute_salary(basicpay, allowances,
deductions);
```

In the earlier statement, "call" is a keyword to call a subroutine. Most programming languages do not use this kind of keyword. Just the name of the subroutine is sufficient to call the subroutine. "compute_salary" is the name of the subroutine being called. The three words, "basicpay," "allowances," and "deductions" are the arguments being passed on to the subroutine. "empsalary" is the variable to which the value computed by the subroutine is assigned.

Now somewhere in the program is our subroutine named compute_salary. It would look something like this:

```
sub compute_salary(a, b, c);
int a;
float b;
float c;
float d;
/* processing statements */
return d;
```

The word "sub" is the keyword to denote that this statement declares a subroutine. Some programming languages require such a keyword in front of the name of the subroutine, but some programming languages do not require such a tag. The variables in the parenthesis are the values that receive the values passed on by the calling program. They have the same number as those in the calling program. Their type also needs to be the same. The variable "d" in the subroutine is the value into which the result of the processing is stored. The statement "return d" causes the computed value to be returned to the calling program, and it would be received by the variable "empsalary" declared in the calling program. Summarizing this discussion, let us enumerate the rules of argument passing:

1. The number of values passed from the calling program to the subroutine must be the same as those of the variables mentioned in the parenthesis of the subroutine being called.
2. The order of the arguments passed from the calling program must be the same as the order of the receiving variables declared in the subroutine.
3. The data type of each of the arguments being passed on by the calling program must be the same as the corresponding receiving variable declared in the subroutine.
4. The size of each receiving variable in the subroutine must be equal to or greater than the corresponding argument in the calling program. If the size of the receiving variable is smaller than that of the corresponding argument, then the value may be truncated and the result can be inaccurate.

5. The names in the arguments (variables supplied by the calling program) need not be the same as those in the parameters (variables in the subroutine for receiving values from the calling program).

6. The subroutine can return one or multiple values back to the calling program as required by the situation at hand.

7. The size of the variable that is designated for receiving the value returned by the subroutine must be equal to or greater than the corresponding variable of the subroutine. If the receiving variable is smaller in size than the returning variable, the value may be truncated.

Message Passing

OOM uses the term "message passing" for communication between objects. The C programming language family does not have any programs or subprograms. They just have functions, one of which must be the "main" function. In OOM, all are objects. Therefore, there is no calling program or a subroutine. But objects can and do call each other. When all are treated equally, the phrase "parameter passing" does not look appropriate. So, they used this phrase. In our terminology, there is a calling program and there is a called program or, in OOM terminology, there is an object calling another object and there is an object receiving that call and returning the result of processing. My averment is that the called program (or object) is a subroutine of the calling program (or object). My definition (or that of Mr. Dennis Ritchie, who developed the C programming language) of a subroutine is that it is a program that cannot be converted into an executable program on its own. The message being passed between the objects usually contains:

1. *The name of the object being called*: In a program, many objects are likely to be called. Therefore, the message needs to contain the object ID so that the message can be passed on to the appropriate object.

2. *Function ID*: An object is likely to contain multiple functions within it, so the message needs to contain the function ID so that the right function is called into execution.

3. *Information*: Information consists of data items that need to be passed on to the function in the object that is being called.

The results are returned by the function using the same mechanism of message passing.

Advantages of Using Subroutines

1. *Reduction in complexity in the main program, making it easier to understand*: If we write a long program, it becomes very difficult to understand for the purpose of quality control, debugging, or maintenance. With long programs, people need to

remember multiple data items, multiple processing tasks, their interrelation, and so on. To be able to do so, we need super skilled programmers who need to be paid higher rates, and they are in short supply! With shorter programs, programmers can easily understand and debug or modify easily. Normal programmers can easily understand shorter programs as they contain fewer actions and fewer data items, making it amenable to effortless understanding. Normal people are less costly and are in greater supply.

2. *Avoid redundancy/duplication of code in the program*: Often times, we need to write the same code multiple times in the programs to cater to common tasks. If we do not use subroutines, we need to insert the same code at multiple locations in the program or application. If we have the same code at multiple locations, we may forget to modify the code at all location when a change occurs. This gives rise to code integrity issues. With subroutines, we can avoid this code integrity issue, especially during software maintenance. Subroutines can be reused in the program as required.

3. *Use the subroutine across multiple programs*: It is a common occurrence that the same task needs to be performed in multiple programs. If we do not use subroutines, we need to insert the code in all the programs that need the task. With subroutines, we can avoid such a situation. We can code the subroutine once and then link it with the programs that need the task. If we change the code in the subroutine, all the programs using it will automatically get the updated functionality. With this facility, we can effectively reuse the code and reduce the development time of applications.

4. *Build libraries*: Besides the previous benefits, subroutines facilitate building libraries, especially in the modern programming languages. We can include all the common subroutines into one library and build it as a library. Then, any program needing a subroutine that is included in the library can use it and build an executable by linking with the library at the link time during the process of building the executable file. In the case of DLL (Dynamic Link Library), the library will be linked at the runtime. It promotes code reuse and reduces the total software development cycle time, too.

5. *Make it easier for debugging during initial development and software maintenance during production*: Smaller programs are easier to understand, making it easier to debug them during the time of initial development and during software maintenance in the production runs. The time taken for debugging or maintenance is not linear in proportion to its length. If a 50-line program takes one hour to locate and make a modification, either for fixing a bug or maintenance, a 1000-line single program takes much more than 20 hours ($1 \times 1000/50$)! But if we have twenty 50-line programs, we take just 20 hours to carry out the same maintenance task! The same is true in the case of initial development too. Complexity increases when the quantity or size increases. Imagine posting a single letter and posting 10,000 letters—a single letter in an envelope takes practically no time at all! But posting 10,000 letters and inserting the letters in the envelopes can take days if you do not use a machine. In case you use a machine, see—the complexity has increased so much that you needed a machine for inserting letters in envelopes! So, we need to keep the programs shorter, and subroutines are a great mechanism to do so.

6. *Make quality control and testing easier and quicker*: When you walk through a program, you need to remember most of the code to determine the action being taken by the program. If the program is long, our memory fails us, and we need to walk back and forth through the program to understand what the statements are proposing to do. With shorter programs, our memory can easily handle the code, and we do not spend time walking back and forth during reviews. Similarly, during testing, too, we need to run fewer steps to test a short program. We need to run disproportionately more steps to test a longer program. Shorter programs increase the productivity of our quality-control activities and reduce the time spent on quality control of our programs. Subroutines are a great way to do so.

Best Practices in Programming Subroutines

Here are the best practices in developing subroutines:

1. *Limit the tasks to one*: No programming language places any restrictions on the number of tasks that can be included in a subroutine. Usually, it is a best practice to limit the subroutine to one task. Of course, occasionally we need to include more tasks in a subroutine, especially when these tasks are closely related and are selected by clicking an option. But we need to see if we can have a separate subroutine for each task even in such cases. When we limit the subroutine to one task, we can generalize it and reuse the code in other projects or applications.

2. *Limit the code to 50 lines*: Again, no programming language places any restriction on the number of lines a subroutine can have. Since the very purpose of a subroutine is to shorten the calling program, coding a long routine defeats the very purpose of a subroutine. But, the question always faced by programmers is—how long is really long? I have seen some organizations defining 50 lines of code as a long program, especially for subroutines. It is easy for programmers as well as the reviewers to hold 50 lines in memory. Of course, this is an arbitrary number that I observed to be serving the purpose well. What I advocate is that each organization ought to define its own maximum length of a subroutine for adherence in that organization. The programming language has an impact on the number of lines that can be easily understood for easy programming and reviewing. So, the definition of the maximum number of lines of code that can be permitted in a subroutine ought to be different for each of the programming languages used in the organization.

3. *Make it universal*: Remove hard-coding and make the subroutine universal so it can be used by other programmers. I would say that all subroutines must be written in such a way that they can be called from any program needing that task to be carried out. To achieve this objective, we should not use any static variables, either defined inside the subroutine or defined in the calling program. We should completely avoid manipulating the variables declared in the calling program. We should receive all parameters and return all the results back to the calling program. We should declare all intermediate variables as required within the subroutine itself. This way, we can hive off the subroutine and make it part of a library of subroutines at the organizational level for use by all the programmers and projects inside the organization.

Pitfalls in Programming Subroutines

Here are some of the pitfalls that I have observed programmers falling into while coding subroutines:

1. *Making it too long*: This is one common pitfall that the programmers often fall into. We get carried away and make the subroutine too long. When we come across a situation in which may need a subroutine that needs to be longer than 50 lines, then we need to consider the possibility of splitting the subroutine into two or more subroutines. Especially in scientific and mathematical programming, we may need longer subprograms. In such cases, we can have a separate specification of subroutine length if we cannot split the processing into multiple subroutines. My suggestion is to treat 50 lines as the optimum length for a subroutine. Any subroutine longer than 50 lines is too long a subroutine. Do not argue—how about 55 lines! Any specification has certain tolerance to it. Generally, 5% tolerance is accepted in most cases. In some cases, a 10% tolerance is also accepted. For a subroutine, if we accept a 10% tolerance, it can vary up to 55 lines on the higher side. But even in such cases, we need to see if there is a possibility to hive of some of the code into another subroutine.

2. *Stuffing multiple tasks*: This is another trap we often fall into. Instead of limiting the subroutine to one task, we are tempted to stuff more tasks into one subroutine, especially when the tasks are small steps needing very few lines of code. No organization would place any restriction on the minimum number of lines of code a subroutine must have. Therefore, we should not be constrained to include more tasks in a subroutine because it is too short! As far as possible, we should limit the number of tasks in a subroutine to one.

3. *Calling of other subroutines leading to deadlocks*: While developing subroutines, we may need to include code to call other subroutines. When we call a subroutine, we need to ensure that the called subroutine avoids calling the calling subroutine! This causes a deadlock as the subroutines would be calling each other. In some cases, we may call a subroutine that calls another subroutine which calls the present subroutine! I will explain:

   a. Let us assume we are developing a subroutine named subroutine-1.

   b. Subroutine-1 calls subroutine-2.

   c. Subroutine-2 calls subroutine-3.

   d. Subroutine-3 calls subroutine-1.

   e. As you can see, this enters into a deadlock and freezes the computer or the program! We should avoid this kind of calls for subroutine.

All in all, subroutines are a great way to reduce the size of programs and thus limit their complexity to manageable levels. We ought to and actually are making extensive of use of subroutines in the software industry.

16

Building and Using Libraries

Introduction

When we work in organizations or have our own software development set-up, we soon realize that many of the functions recur in our software development projects. It is possible that these common functions can have minor deviations from each other, but most of the code and functionality remains the same. One way is to write the code, copy it to the new projects, and make necessary modifications as needed. This method is not seen as a very professional one, as it takes time not only to make modifications and repeating the code but also for the vital quality-control activities of code review and testing. The second method is to build libraries of useful subroutines for commonly used functionalities, and then use them when developing software in different projects. This aspect is discussed in this chapter.

Types of Libraries

A normal library is filled with a collection of different books. A library in the context of software development refers to a collection of independent subroutines that can be called by other programs. In fact, every programming language provides a set of libraries along with its development kit. When we run the step of linking during the process of preparing the executable program, the object code of the relevant subroutines from these libraries are attached to the program code we wrote and then the executable file will be prepared. The COBOL language provides for copy books, that is, files that will be brought into the executable file during the process of preparation of the executable file. A copy book is a file containing COBOL code and we can bring it in by inserting a COPY statement in the code. In other programming languages, we simply use the routines inside the program and include the library during the linking step, along with other libraries, during the process of preparing the executable file. In the present-day IDEs (Integrated Development Environments), we include the name of our library in the project reference. How exactly we include our library along with the libraries supplied by the SDK (Software Development Kit) supplier is a matter of detail and changes from one development platform to another.

There are three types of libraries:

1. *Static libraries*: This is the initial type of library to be used in the third-generation programming languages. It began with the FORTRAN programming language, which supplied a large set of mathematical routines in its libraries. The mathematical

library was the reason why the FORTRAN programming language was selected for mathematical and scientific programming and is still the number one choice for that type of programming, even today. In this kind of library, the library code is attached to the program code, increasing the size of the executable file. Initially, the entire library code was attached to the executable file irrespective of the number of routines from the library that were used in the program. Later on, this was changed to attaching only those routines that were actually used by the program. This reduced the size of the executable file. Still, the size of the executable file was larger than the program size to the extent of the subroutines included in the program. In those days of constraints on the available RAM, increased size meant increased pressure on the resources available, leading to degrading the performance and throughput of the computer. There was one major advantage with static libraries and that was the program was one executable! All the required code was inside the executable file. It can be easily carried to another identical computer with ease.

2. *Dynamic libraries*: Dynamic libraries are not attached to the executable file during the linking stage but are needed on the disk for loading into the RAM during the execution of the program. The dynamic library may be loaded into the RAM either at the time of starting the execution of the program or during the first time it is called by the program in execution. The library needs to be available on the disk in the search path set for the execution of the program. This over time developed into the DLL (Dynamic Link Library) that is being extensively used in the software industry today. DLLs stay resident on the disk and are loaded into the RAM as and when called by a program in execution. It is cleared from the RAM when the called program is closed along with the other resources held by that program. What happens when more than one program calls for the same DLL? It will be loaded into the program space of both the programs. In other words, two copies of the DLL are loaded into the RAM but at different locations. Each executing program would have its own copy of the DLL in its space in the RAM. Presently, when a routine in the library is called by a program in execution, only the routine would be loaded into the RAM. The entire DLL would not be loaded into the RAM.

3. *Shared libraries*: A shared library is also referred to as a runtime library. We have noted the disadvantage of the DLL in the previous section. Some routines are needed by many programs in execution, and loading a separate copy for each would consume significant amount of space in the RAM. This is common in the software that constitutes the OS. So, the OS developers have developed a technique to keep just one copy of the library in the RAM and then allow other programs in execution to use it a needed. Instead of loading the entire executable code of the routine for each of the calling programs, only the data needed for the variables is created separately for each of the programs using the library routine. It, in fact, mimics the multi-user OS. In multi-user OS, each user would be provided with a session space in the RAM and the program remains one. Shared libraries also work in the similar manner. Implementation of shared libraries is in the domain of the OS. Unless the OS provides this facility, we cannot use shared libraries. Many OS use shared libraries. Most multi-user OS provides facility for using shared libraries developed by application programmers.

As you can see, the difference is in the usage of the library. All are built the same way. Now let us see how the libraries are built so that we can use them in our software development projects.

Building Libraries

The components that can be included in a library are object programs. An object program is the compiled version of the source code. We cannot usually include GUI controls in a library. Most computers provide separate facilities for building libraries with the GUI controls. Such libraries are usually referred to as visual components libraries or a name similar to it. Usually we build libraries that perform some processing functions using program statements. Different programming languages and OS have different facilities for including routines in the library. The beauty of a library is it can be used with programs of any language. You can write the code of the routine included in one language but call it in a program being written in a totally different language. For example, the code in the library routine might have been written using Visual Basic language and once it is built into a library, it can be called by a program written in Java!

Now here are the steps in building a library:

1. Write the routine using the source code. It could be written using a text editor or an IDE (Integrated Development Environment). The routine should be written such that it does not call for routines from any other code libraries including the libraries supplied along with the SDK. It should be a stand-alone program not needing any external routine for its functioning.

2. Compile the routine to make it into object code. All development platforms provide this facility to compile any routine into its object code. Remember that an executable file is primarily object code linked with the required libraries.

3. Develop all the routines you propose to include in the library as stated in step 1 earlier. Then compile all those routines into their corresponding object files.

4. Build all those routines into a library:

    a. All IDEs provide a facility to build a library.

    b. When you select the option to build a library, you need to specify all the routines that are intended to be included in the library. You also need to give a name with which the new library is accessed. All these names have to adhere to the rules of naming specified by the IDE.

    c. The IDE builds the library and assigns it the name and stores it in the directory specified by you.

    d. In some cases, you may have to issue an appropriate command to build the library from the command prompt and supply the names of the object files being included in the library as well as the name for the new library.

5. Now the library is ready for use!

Usually, all IDEs provide facilities to build new libraries as well as to add new routines to an existing library. We need to select the option of either to create a new library or to add routines to an existing library. Then the IDE will carry out the command and build a new library or add routines to an existing directory as specified by you. If you are building a library from the command prompt, you need to specify an option to either create a new library or to add routines to an existing directory. These options need to be specified, adhering to the rules of the command keyword of the OS.

The process of building a DLL is also similar to the steps described here. You will be using the facility provided in the IDE to build a DLL.

In fact, the process of building a library is simple and its benefits are huge! But alas, I do not find many programmers or organizations making use of this great facility.

How to Use Libraries

To use libraries, we need to include the newly built library at the linking stage along with the standard libraries. In the bygone days, the programmer performed the steps of compiling and linking in two steps, but later on, the suppliers of the programming languages began combining these two steps into one and provided a single command that takes the source code and performs all the steps of converting the source code into object code, then link it with the libraries supplied along with the programming language, perform any other steps necessary, and deliver the executable file. All this with a single command issued from the command prompt. Then came the IDEs which reduced the effort of the programmer even further! So, today's programmers may not recognize the steps involved in building an executable file beginning with a source code file! When the command to build the executable file is issued from the command prompt, it automatically includes all the libraries supplied along with the programming language for linking the object code. But those suppliers of programming languages recognized the need to provide an option to include other libraries not supplied by them. They provided options to include our libraries in the command line. We need to specify the names of our libraries along with their location in the command line, adhering to the rules of the command keyword provided by the programming language or the OS.

When we compile a program that used our libraries, we need to remember that the references would be shown as syntax errors! That is because, the compiler resolves all references from the keywords that form part of the programming language or its libraries and the variable names declared within the program. But the names of the routines from our libraries, including the variables declared therein, would not be resolved, throwing up syntax errors. Once we link the program with our libraries, these errors would vanish. Usually the command supplied along with the programming language provides an option to include our libraries too during compile stage so that these errors can be avoided.

With the present practice of using IDEs, we need to include the names of our libraries in the list of project references or list of libraries. Each IDE is unique, and we need to find the appropriate option and include the names along with their location of our libraries. Once we do this, the IDE would resolve the references in the program and prevent errors. Once we include our libraries in the Project References option of the IDE, it would include them in the installation package automatically. If you are using a specialized tool like the InstallShield to prepare the installation package, it would also automatically include all the libraries included in the project references section of the IDE.

We need to note here that different OS and development platforms provide different facilities for including libraries at link time, and we need to learn the same and use our libraries accordingly. I have seen some OS that allow libraries to be included at the time of execution too. Especially in the Internet applications, the execution is in interpreter form. That is, the source code statements are not precompiled into object code. Each statement is compiled on the fly during execution, then linked to the relevant libraries and executed. If the execution encounters an error, the OS throws up an error and aborts the execution of the program. In such cases, the libraries are linked to the object code during the time of program execution.

Document the Routines in the Library

One aspect in using the libraries is the technical aspect of how to include the library to prepare the executable. The other is to provide information to the programmers that the required function is already available in a library and that it does not have to be coded again. For this we need meticulous documentation of each routine in the library and a mechanism for indexing the library routines such that the required functionality can be quickly located.

Each routine needs to be documented. I suggest the following aspects to be documented:

1. The name of the routine.
2. The function achieved by the routine in detail.
3. The algorithm used by the routine—either diagrammatically, descriptively, or both as necessary.
4. The parameters to be passed on to the routine, including their size and type.
5. The results being returned, including their size and type.
6. Any other relevant aspect depending on the situation at hand.

Once each routine is documented in this manner, we need to index it in the table that holds the searchable index of all the libraries and the routines therein. I suggest the following structure for such table:

1. Name of the routine.
2. The name of the library in which it is available.
3. The main function achieved by the routine in brief. This description gives the general description to the person searching for the right routine so the person can access its document for more details.
4. The name and location of the document in which the functionality of the routine is fully described.
5. Any other relevant details.

With such a mechanism, the libraries can be effectively utilized by the programmers and organizational productivity can be greatly increased.

Organizational Role in Building and Using Libraries

While it is the programmers that do all the work in building and using libraries, it is the organization that facilitates the work. By building and using libraries, it is the organization more than the individual programmers that derives the real benefit. Programmers are professional workers putting in eight hours per day. During those eight hours, they write whatever code that is assigned to them. By reducing the amount of code to be produced by reusing the code already developed and tested, the organization saves the effort and therefore money, increasing the ROI (Return on Investment) as well as the profit. Therefore, the organization needs to focus attention on building libraries and making them available to programmers for use. Here are the activities that need to be performed by the organizational managers in this crucial function:

1. Provide funding and resources for building libraries including:
   a. Identifying the routines that can be included in a library.
   b. Modifying them to make them flexible for use in different scenarios.
   c. Arranging for documenting the identified routines so others can make use of them.
   d. Building libraries.
   e. Building a repository of all the libraries built in the organization and maintaining it regularly.
2. Provide funding and resources for maintaining the libraries:
   a. Add new routines into libraries.
   b. Rectify defects, if any, that surface later on in the routines included in the library.
   c. Improve the routines for better efficiency or to make use of newer developments in the technology.
3. Provide funding and resources for building facilities to search and locate the needed routine:
   a. Build a database to hold the data of all routines and libraries available in the organization.
   b. Assign the responsibility to some agency to maintain this database.
   c. Develop the necessary programs to easily search and quickly locate the needed routine.
   d. Assign the responsibility of maintaining this software for searching and locating the routines as necessary.
4. Provide funding and resources for encouraging the use of libraries as well as monitor the results obtained by building and using the libraries in the organization:
   a. Assign the responsibility to some agency that can be consulted by programmers and project leaders as necessary in the effective use of the routines in the libraries.

    b. Advise the software development teams as necessary to encourage the use of libraries in the organization.

    c. Champion the cause of developing and using the libraries in the organization.

5. Any other relevant activity to encourage and promote the use of libraries in the organization.

All of the above constitute the components of an organizational framework that facilitates and encourages the use of libraries in the organization.

Now, the individual programmers also have a role to help the organization maintain a healthy environment for building and using the libraries. These are:

1. Identify the routine they came across during their work that can be included in a library and inform the organizational agency responsible for maintaining the libraries for possible inclusion.

2. Assist the agency responsible for maintaining the libraries in their assessment of the suitability of the routine for inclusion.

3. Assist the organizational agency in documenting the routine for future use.

4. Identify any opportunities for improvement in any of the aspects of maintaining the organizational libraries for more effectiveness.

5. Identify any opportunities for improvement, including the defects that surfaced in the routines that are in a library, and assist the organization to replace the existing routine with the improved routine in the library.

6. Any other activity that is necessary and assigned by the organizational agency responsible for maintaining the libraries in the effective and efficient maintenance of the building and using of libraries, including training newer resources in the organization.

In this way, the organization and the individual programmers need to work shoulder-to-shoulder in a close-knit manner to derive benefits from the activity of building and using the libraries. While the organization derives financial benefit, the individuals derive the benefit of avoiding the monotony of writing the same code again and again.

17

Programming Device Drivers

Introduction

Device drivers are programs that are developed to interface between the computer and the device or machine connected to the computer and assists the computer in controlling such devices or machines. A device can be a printer, a VDU (Visual Display Unit), a tape drive, a mouse, a camera, a CNC (Computer Numerically Controlled) machine, an airplane, a rocket, a car, or any other such machine. Programming is not limited to computerizing business operations, developing decision support systems, or mathematical solutions. Computers have wide-ranging applications. In all the applications, the output has to finally be delivered on a machine or device such as a printer or another machine. Computers are controlling nuclear reactors, rockets, and flow process production systems such as fertilizer and pharmaceutical manufacturing. In a business environment, when we purchase a computer with its peripherals, we get it with system software and the needed device drivers that are supplied along with the computer. The supplier of the device also supplies the software or the device driver that controls the device supplied by their organization. In this chapter, let us discuss how to develop the device drivers. However, this chapter provides a brief introduction so that you can build on it and develop programs needed to interface and control the devices. It is by no means intended to make you an expert device driver developer.

What Is a Device?

From the standpoint of a programmer, what is a device? For our context, a device is any machine or gadget that can be controlled by a computer. A programmer knows how to write a computer program that can process information but not run a device. A computer program can deliver the processed information to any output device, but cannot pilot an airplane or print an output on a paper. Those activities such as printing, piloting a plane or a rocket, or running a machine have to be performed by the device itself. What a computer can do is pass on such information that is needed by the device to run itself effectively. In the absence of a printer, a typist used to print the matter on paper using a typewriter machine. The airplane was piloted by a qualified and certified pilot. The human being tending to the machine performed two functions, namely, running the machine and making informed decisions. Running the typewriter involves loading the paper, pressing on the keys with the right amount of force, return the carriage of the typewriter to the beginning position when the end of the line has reached, formatting the output by setting tabs appropriately, and so on. The decisions the human being made included the positions where to set the tabs, when to return the carriage, ejecting the paper when it is filled up and loading the new paper, transferring the information from a handwritten note to the paper, and so on. Engineers have come up with gadgets that can take over the actions the human being performed, and the computer took over the decision-making portion of the human being. Our device driver software makes the decisions to drive the device as well as to transfer the data required by the device to implement the decision. In other words, the device driver software gives commands to the device along with the data to implement the command.

Each device that is designed for working under the control of a computer has some interfacing hardware to interact with the computer. This hardware would perform the following functions:

1. It interacts with all the internal hardware of the device.
2. It interfaces with the computer by performing a two-way communication with the computer.
3. It resolves the commands received from the computer, determines the subassembly to which it is intended, and delivers the command along with the data to the selected subassembly.
4. It provides appropriate responses to all the commands received from the computer.
5. It raises error-condition alarms to the computer.

Each of the subassemblies of the device would have a means to perform the work as commanded by its computer interfacing component.

In short, our programs interact only with this interfacing hardware and software component of the device. We need not concern ourselves with how the device implements our commands and delivers the right output expected of it. For us programmers, our vision of the device is limited to providing the right command with the right data at the right time to the interfacing component of the device.

Functions Performed by a Device Driver

The device driver software performs several functions depending upon the device. The commands and data provided by a device driver controlling an airplane would be vastly different from those of the device driver that becomes the radio jockey playing the music demanded by their customers. But there is some common thread in all of the device drivers. We can basically classify the functions performed by the device driver software into two classes:

1. Core functionality actions
2. Ancillary functionality actions

Core functionality actions are those that deliver the expected output from the device. A printer is expected to print the output on paper. A music player is expected to play music conforming to the playlist. An airplane is expected to reach its destination flying the specified route.

Ancillary functionality actions are those that keep the core functionality actions safe and secure for the end users by protecting them from failures and malfunctions of the device.

While we cannot generalize the core functionality actions into subclasses, we can generalize the programs we need to develop for the ancillary functionality actions. Let us now discuss about coding these two main classes of actions.

Coding the Core Functionality Actions

Core functionality consists of implementing a set of commands. Each device that is designed for working under the control of a computer comes with a set of native commands, often referred to as the "primitives" of the device. Each primitive requires some data, but it usually has a default value. For example, for the LF (line feed) command in a printer, the paper is pushed up by one line if the command is not accompanied by a number. But if the LF command is followed by a number, let us say, 2, then the paper is pushed up by two lines. Another aspect we need to note is that in the same class of devices, even with the same capacity, the primitives would be different. For example, in the printers made by HP, the command set would be different from those of a similar printer made by Canon! We may be able to reuse some of the code in these two cases, but the differences would be significant.

To achieve the core functionality, our device driver needs to be given a data file by the computer. Using this file, our device driver drives the device by performing the following actions:

1. Our device driver reads the data one set at a time.
2. It translates the data into the commands and data appropriate for the device.
3. It transfers the command and the data to the device.
4. It waits for the acknowledgement from the device.
5. If the acknowledgement received from the device indicates success, the device driver moves to the next item in the data file.
6. If the acknowledgement received indicates failure, then the device driver resends the command and data to the device once again. In fact, the device driver continues to send the command and data until a success acknowledgement is received or a pre-decided number of attempts occur before raising an error-alarm.
7. Even when the command and data are passed on to the device successfully, the operation at the device may fail due to reasons of hardware malfunction, tools getting damaged, or such other error conditions. In such cases, our device driver needs to receive the error interrupt raised by the device and attend to it.
8. When all the data in the data file is exhausted, the core functionality actions are completed, and the finishing routine of the device driver takes over to close the program and flush all the variables.

What commands to give and with what data depends on the device itself, and we need to develop the needed programs accordingly.

Coding the Ancillary Functionality Actions

Ancillary functionality actions are similar in most devices, but they are never identical. Here are some actions that are mostly common to most devices:

1. *POD* (*Power-on-diagnostics*): Every computer-controlled device has built-in software that performs this function. If we are developing that software, yes, we need to program this functionality. However, in the present context, we are talking about a device driver that is resident on the computer and delivers output to the device. POD software of the device contains a functionality to check that every subassembly or component is in working order. To do this, the POD software of the device sends a signal to the component or subassembly, through its hardware interface, asking for its status, and the component or the subassembly responds to it with its status. The status could be healthy and working, an error condition, or no response at all. Then the device flashes a light on its control panel to indicate error with a description of the error or an error code that the human beings can understand or derive its meaning from a user manual. This status is stored in the buffer of the device and is made available to the computer. So, the first action we need to perform in our device driver program is to obtain the status of the device from its interface unit and interpret it. If the device is healthy, we move to the next functionality. If there is an error, then our program needs to display a message to the user, wait for the response, and take appropriate action based on the response. In a nutshell, here are the functions to be programmed in this section:

   a. Send message to the device requesting status.

   b. Interpret the status received from the device.

   c. If the status indicates that the device is busy serving another program, we need to flash a message that the device is busy and await user response. In some cases, like the printer, we can put in the request to queue up our job. In such cases, we need to place our job in the queue of waiting jobs and wait for the interrupt of the device indicating that it is free to take up our job.

   d. If the status indicates healthy functioning of the device, pass on the control to the next section to deliver the output.

   e. If the status indicates an error, interpret the error, retrieve the message associated with the error code, display it on the screen, and await the user response.

   f. The user may rectify the error and ask the program to continue, in which case we need to recheck the status, ensure that the device is healthy, and then pass on the control to the next section.

   g. If the user asks to cancel the action, then our program needs to close all open files or database tables, flush all variables, and end its run.

2. *Issue commands*: Once our device driver software determines that the device is in healthy working condition, it needs to issue a series of commands to the device. Each of these commands trigger some action on the device. These commands range from enquiring status, to POD, to data transfer, to handling error conditions, to releasing the device after our job is completed. Each command of the device needs to follow certain device-specific syntax, and we need to follow those rules strictly.

3. *Data transfer*: Once the device is dedicated to our program, we begin transferring a series of commands followed by corresponding data to the device. We follow these steps in transferring data:

   a. Query the device to ensure that it is ready to receive data.

   b. Begin transferring data, one packet at a time. The size of the packet differs from device to device.

   c. Wait for the acknowledgement from the device that the packet is received and that the data is in healthy condition. If the acknowledgement is not received, the program needs to resend the packet once again. We repeat sending the packet until our program receives an acknowledgement from the device that the data received is in healthy condition.

   d. Repeat the previous three steps until all the data needed by the command is transferred to the device.

   e. Move to execute the next instruction in the program.

4. *Respond to device interrupts*: The device needs to place interrupts on the CPU of the computer quite a few times. Some occasions would be device-specific. Here are some general reasons which are common to most devices:

   a. *The receiving buffer is full*: Each computer-controlled device has a small amount of RAM, and it gets filled up pretty quickly. In this case, the device places this interrupt on the CPU to halt further data transfer.

   b. *Receiving buffer is empty*: When the data transfer is halted due to buffer being full, the device places another interrupt on the CPU to resume data transfer. Then the CPU would resume transferring the remaining data.

   c. *Error condition*: The device may experience a variety of errors like an output material like paper is exhausted, the tool is not working, there is a paper jam, and so on. Then the device places an interrupt on the CPU to take appropriate action.

   d. *Online*: Sometimes, the communication may break down between the computer and the device for some reason. In such cases, when the communication is reestablished, the device places this interrupt to indicate that it is ready to receive instructions from the computer.

   e. There would be many other occasions that necessitate placing an interrupt on the CPU, and our program needs to be ready to handle all such interrupts.

5. *Error trapping*: This is a major function of the device driver. The device is subject to a host of errors. A device driver software needs to handle all error conditions, trap them, and steer the computer to a smooth passage to other functionality without effecting the computer functioning in any manner:

   a. Buffer full.

   b. Out of paper.

   c. Not powered up.

   d. Error conditions.

   e. Ribbon, cartridge, and ink.

   f. Parameters out of range.

   g. Tool broken.

   h. Process of the device going out of acceptable range.
   i. There would be many such errors specific to each device, and all such errors
       have to be handled.

In this manner, we need to program all the functions required to drive the device. All we
do in programming devices is interact with the digital interface built within the device and
pass appropriate commands along with the needed data in the specified format. The rest
is taken care of by that device interface to produce the expected results from the device.

18

Programming Multi-Language Software

Introduction

Now the world has shrunk; if not in land area, it has shrunk in terms of reach, especially for products and markets. Gone are the days when we develop software in English and expect all others to master the English language if they like to use our products. Others have changed, and we also need to adapt to the ways of the world and develop software in such a manner that the people completely non-experts in English can use our software. The tools for doing so are available right now. Most software development platforms provide facilities to develop software for use by people of different languages.

Why do we need our software to work in any language other than English? Of course, English is the language spoken all over the world. No other language is spoken by so many people as English. But the people who do not speak English in the world outnumber those that speak English. Thanks to the low-cost IBM PC and its clones, computers are now used all over the world. Until recently, we were developing software only in English because the OS was supporting only English. But now, the OS, especially that of the PC, is supporting other languages, including all European, Asian, and Arabic languages. Present-day keyboards allow for the entry of data in native languages other than English. So, if we do not develop software in languages other than English, we will be losing a large chunk of market. Our software is not like a book that can be translated into the native language and released. There is also no point in developing a separate version of software for each of the languages. We need to build in features that allow users of different languages to use our software. It is possible in the present day, and this chapter gives you an insight into developing multi-language software.

I need to make one thing clear before we proceed further: We do not develop software in multiple languages, but we develop software that is amenable for use by people using languages other than English.

Attributes of a Multi-Language Software

What constitutes a multi-language software? A multi-language software has the following attributes from an end user's standpoint:

1. All the prompts for the user on the screen can be displayed in any language, within the set of languages in which it was designed to be used.
2. All the tool tips are displayed in the chosen language.
3. All the help displayed is in the chosen language.
4. All the headings and other labels on the report are generated in the chosen language.
5. The input is allowed to be entered in the chosen language.
6. The end user uses the software in one language with which he/she is comfortable. The end user would not use the software in different languages in different sessions.

But how does a multi-language software look like from the programmer's standpoint? I am not aware if any programming language is available in any language other than English. Efforts were certainly put in to develop programming languages in languages other than English, and perhaps in some countries they may be available. But once the programs are compiled, they will be in machine language that is in zeroes and ones! The source statements are hidden from the end user who sees the software through the screens and reports or the actions of a machine. So, the language of the programming language does not matter to the end user. What we need to achieve the multi-lingual attribute is in the user interface both on the screen and the reports.

Methods of Achieving Multi-Lingual User Interface

First and foremost, we need to avoid hard-coding of text on the labels. Only then can we achieve multi-lingual user interface. We need to assign text to the labels programmatically during the FormLoad event, which is the first event executed while loading any functionality. We enter the labels in a table in the DBMS and load the labels onto the screen or the report, reading them from the database table. Here are the methods to make the software multi-lingual:

1. Make the software amenable for use in one language only but with a customizable user interface
2. Make the software amenable for use in English and one other language
3. Make the software amenable for use in multiple languages

Let us discuss each in detail.

Make the Software Amenable for Use in One Language Only but Customizable

In many business applications, customers would like to use different terminology, especially in the case of COTS products. For example, in most organizations, the supplier of input items is referred to as a vendor, but I had seen some organizations calling that same entity a supplier! Therefore, it would be to our advantage to allow the customer to change the text on the label prompts on the screens and headings on the reports to their liking. We can easily achieve this facet in our programs. We achieve this in this manner:

1. We place all the label text in a database table that has only one field in which the label text is stored.
2. When loading a screen/report on the application, we read the right label text from the database table and assign the text to the labels appropriately.
3. We provide a separate screen in system administration functionality to change the label text as desired by the customer.
4. We take the following precautions for this functionality:
   a. The action of the changing of the label text is allowed only for the system administrator.
   b. The label text change is allowed for the entire functionality only. Each individual user cannot have different label text to suit his/her liking.
   c. The size (the number of characters) of the new label text is restricted to the maximum size of the label.

The term "label" used in the previous bullets includes all text visible to the user including tool tips, captions on buttons, menu items, and so on.

This method as described is keeping in view multi-user software. If it is a single-user software, the method remains the same except that we give the user the facility to change the label text as desired by the user.

Make the Software Amenable for Use in English and One Other Language

We use to this method when we develop software for a customer who does not understand English. In these days of globalization, all countries are using computer-based systems for most business as well as government applications. They would like to use their own language for their information processing needs. Most countries are outsourcing their software development work to countries like USA and India who use English as the primary language for software. In such cases, this method is utilized to serve the customer and at the same time does not put any pressure on the programmers.

In this method, we achieve the objective in the following manner:

1. We design a database table that has two fields. One field would contain the label text in English, and the second field would contain label text in the desired language.

2. We develop the software in our usual language, that is, English, storing all the label text in the database table field specified for English language label text.

3. Concurrently, we get someone to translate the label text to the desired language and store it in the database table field that is designated for the desired language label text. Here the size of the new label text in another language needs to be restricted to the size of the original label text placed on the screen or report.

4. We will also have a facility to capture the preferred language of the user and store it in the configuration file. When we launch the screen/report, we read the preferred language and then read the label text from the appropriate field and assign it to the labels as necessary. This will achieve the objective of showing the user interfaces with the language desired by the user.

5. We also build in a facility to change the label text as desired by the end users to suit their needs. However, this facility is restricted to be accessed only by the system administrator in the case of multi-user software. No such restriction would be placed in case of single user software.

6. The size of the new label text when the original label text is changed needs to be restricted to the size of the original label text of the label placed on the screen or report.

The term "label" used in the earlier bullets includes all text visible to the user including tool tips, captions on buttons, menu items, and so on.

In fact, this can serve the purpose of providing software in the language desired by the customer. However, we need to know who the customer is before we sell the software and be allowed time to translate the label text. This situation is possible in custom software development. This is not a good alternative for the COTS product scenario.

Make the Software Amenable for Use in Multiple Languages

In a COTS product scenario, we do not know who the customer and the desired language are going to be! We place the CD/DVD in stores or make our software downloadable from our website and the customer purchases the software from a store or downloads it from our website from anywhere in the world. The customer must be able to use the software by just specifying the desired language. COTS products are usually single-user versions, and the customer cannot be expected to know English well enough to translate the user interface into their desired language. The user may request to change the label text in the desired language to suit his/her needs. So, in this case, we need to provide the facility to provide the label text in as many languages as the number of different language groups in which we intend to sell our product. Here is how we do it:

1. We design a database table that has as many fields as the number of languages in which we wish to make our software available. One field would contain the label text in English and each of the other fields would contain the label text in one desired language.

2. We develop the software in our usual language, that is, English, storing all the label text in the database table field specified for English language label text.

3. Concurrently, we get language experts to translate the label text to the desired languages and store it in the database table fields designated for the respective languages. Here the size of the new label text in languages other than English needs to be restricted to the size of the original label text of the size of the label placed on the screen or report.

4. We will also have a facility to capture the preferred language of the user and store it in the configuration file. When we launch the screen/report, we read the preferred language and then read the label text from the appropriate field and assign it to the labels as necessary. This will achieve the objective of showing the user interfaces with the language desired by the user.

5. We also build in a facility to change the label text as desired by the end users to suit their needs. However, this facility is restricted to be accessed only by the system administrator in the case of multi-user software. No such restriction would be placed in case of single-user software.

6. The size of the new label text when the original label text is changed needs to be restricted to the size of the original label text of the label placed on the screen or report.

The term "label" used in the earlier bullets includes all text visible to the user including tool tips, captions on buttons, menu items, and so on.

There is one more method of achieving this functionality. In this method, instead of using only one table for all the label text, we use a separate table for each of the languages in which we intend to make our software available. In the configuration file, we capture the desired language and then use the corresponding table while loading the label and other text.

Limitations

Now that we know how to develop software for use in languages other than English, we need to learn the limitations of such development. Here are some such limitations:

1. For this facility to work, the OS of the computer needs to support the language desired by the customer. If the OS user interface is in English, and the user likes to use our software in, let us say, German, which has special character like "ŭ," it would not be possible unless the OS supports it. That is, our software works on the OS and if the OS does not provide support for multiple languages, our software cannot support multiple languages.

2. The European languages are similar to English, but some languages do have much longer spellings than English. So, the label size has to be larger than that which is needed for English. This can be circumvented by storing the label size also in the database table containing the label text of different languages to help accommodate more characters in the label text.

3. In languages of the Arabic family, Indian language family, Chinese, Japanese, and such other languages, it may be difficult to make our software to be amenable for use in those languages because I am not sure if any OS supports all those languages. Fonts are made available in those languages in Windows OS, and perhaps we can use that facility, but I am not sure about other computers, especially the mainframe and mid-range computers.

Of course, I have not given a detailed algorithm for each of these alternatives. I have shown you the way and explained the method so you can design your algorithm. You can use a loop for assigning the label text to labels beginning with the first control to the last control on the screen. In a report, you need to pass values of the label text to the labels on the report as parameters to the program calling the report from the report-generator engine, if you are using one. If you are creating the report programmatically, you just need to assign the value of the label text to the variable used to print the labels on the report.

19

Programming Languages and Their Evolution

Introduction

As programmers, we need to understand the evolution of programming languages from the beginning so we can be prepared when the new programming languages come on to the scene. If you ask me whether learning this history is essential, I would say, "No." But learning the history would be advantageous in that it would prepare you for the next big change. As it has been said, "change is the only permanent thing in this world!" I am presenting here just a gist of the evolution of the programming languages through the generations to acquaint you with the development.

Evolution of Programming Languages

The first computer, ENIAC (Electronic Numerical Integrator and Calculator), was programmed using its hardware. It involved setting switches and changing the wiring. Feeding a new program took a significant amount of time in those days. The Harvard Mark-I (Automatic Sequence Controlled Calculator) used instructions punched on a paper tape. The instructions could be called the program, and they were presented in the sequence of execution. Those instructions were zeroes and ones (0 and 1). We need to understand that, even today, computers can process machine instructions (zeroes and ones) only. What has been achieved is that we have developed translators that can take our high-level language instructions and convert them into machine language instructions. Programming the computers by manipulating the hardware was referred to as 1st generation of programming languages, but it was so christened during the advent of the 3GLs (3rd Generation Languages) that will be explained in the subsequent sections.

While this is clear, the rest of the history of programming languages becomes a little clouded. Assembler languages and FORTRAN (FORmula TRANslator) were developed around approximately the same time. Assembly languages, among which the Autocoder of IBM was most popular, created Opcodes (Operation Codes), which were at a slightly higher level than machine language, which was just zeroes and ones. An instruction in the assembler languages had two parts: the opcode and the operand. The opcode was the instruction, and the operand provided the data to be processed by the opcode. It was focused equally on processing the data and on operating the CPU. It was not easy writing those programs. Life has become easier now in programming computers. There were quite a few assembly languages in those days. Assembly language programs were converted to machine language instructions using what was referred to as an "Assembler" that translated each of the assembly language instructions into machine instructions. These assembler languages were christened as 2GLs (2nd Generation Languages), as they are one level above machine languages. Even today, some portions of OS continue to be written in the assembly language native to the machine!

High-level languages, as they were called, focused only on the procedure for processing the data. When the programmers coded the programs in high-level languages, the compiler inserted the necessary CPU instructions during the compilation process. The coding of CPU instructions was automated, and the programmers were relived of this aspect. When the compiler compiled a high-level language program instruction, it translated the instruction into several CPU instructions as needed to achieve the processing specified by the high-level language instruction.

FORTRAN, which was made available commercially in 1956, is generally accepted as the first high-level programming language. It was from the beginning aimed at solving mathematical problems used for scientific purposes, and it continues to be the leader in that field even today. Of course, the present-day FORTRAN programming language, while it retains the original flavor, is much more advanced than the original set. The FORTRAN programming language focused on solving mathematical problems. It had a large library of routines for solving mathematical problems. To this day, it remains the first choice when it comes to applications that involve processing complicated mathematical problems, which include weather forecasting, astronomical science applications, and so on.

IBM computer 1401 used the Autocoder as its main programming language. It was used for a considerable amount of time, well into the 1970s. We cannot move forward on our discussion without mentioning the legendary Ms. Grace Hopper along with CODASYL. It was Ms. Hopper who firmly believed that computer programs could be written in plain English instead of in the machine language or assembly language. She developed the first compiled language, FLOW-MATIC, for the Univac-I computer. She advocated for the development of a programming language that was computer-independent and could be used on any computer.

Conference or Committee on Data System Languages

The Department of Defense of the USA set up CODASYL (Conference or Committee on Data System Languages) in 1959. Its aim was to develop a standard for the development of programming languages that can be used on multiple computers. It set up four committees:

1. *Executive committee*: This committee set the policies and provided overall supervision of all other committees. It had reviewed the functioning of the rest of the committees and accorded final approval for all the standards published by them.
2. *Programming languages committee*: This committee developed the specifications for a programming language that facilitates exchange of programs and data from one computer to the other. They came up with the COBOL (Common Business Oriented Language) programming language.

3. *Systems committee*: This committee focused on developing the overall framework of data processing. They came up with what was later referred to as systems analysis and design.

4. *Data description language committee*: The purpose of this committee was to define the specifications for administering the data and the systems. The data administration specifications led to the development of the DBMS systems.

Ms. Grace Hopper participated in the Programming Languages Committee to guide them in creating a computer-independent programming language. The committee was influenced by her idea of a programming language that is akin to normal English. The compiler should translate it to the machine language. This led to the development of the COBOL language, which was an extension of her FLOW-MATIC programming language. IBM also influenced COBOL by borrowing ideas from their FORTRAN clone COMTRAN's programming language. The USA's government standardized COBOL as the programming language for all their business data processing applications, which resulted in COBOL being the most popular programming language for data processing applications in the USA and the rest of the world.

The FORTRAN and COBOL languages held sway over the programming and data processing fraternity for a long time, until the 1970s.

The next language that gained significant support from the industry was BASIC (Beginner's All-Purpose Symbolic Instruction Code) developed at the Dartmouth College of New Hampshire by John G. Kemeny and Thomas E. Kurtz. While it was intended for developing programs for the college, its popularity increased due to its simplicity. DEC (Digital Equipment Corporation) implemented it on their PDP series of lower-cost minicomputers. They extended the original BASIC and made it an alternative to COBOL and FORTRAN. BASIC borrowed ideas from FORTRAN. The BASIC programming language introduced a new method of executing the programs, which was referred to as "Interpreting." Instead of compiling the entire program at one time and creating an object program that can be executed on computer, the BASIC language took the source program statements at the time of execution, compiled the statement on the fly, and then executed it. Microcomputers, beginning with Apple computers and later IBM's PC, adopted this language as their main programming language. Microsoft took this to a new level and developed it further—so much so that it is now one of the leading current programming languages under the title of Visual Basic, or VB in short.

Then, during the development of the UNIX OS, the C language was developed at the Bell Labs during the early 1970s by Dennis Ritchie. It went on to become one of the most popular programming languages. The specialty of the C language was that it provided language constructs to access locations in the RAM and manipulate it along with the constructs for normal data processing facilities. In a way, it provided most of the facilities available in assembler languages as well as the features of the higher-level programming languages. Now, there are quite a few languages of its ilk, so much so that they are referred to as the C-family programming languages. C++ and Java belong to this family. There are many vendors that supply these language compilers to organizations for developing programs.

The Pascal programming language was developed by Professor Nicklaus Wirth, who released it in 1970. It was named after Blaise Pascal, the renowned French scientist. The language had the influence of the ALGOL language on it. It was a parallel development to C, and both had similar objectives and had some similarities in their syntax, too, especially in the statement blocks and the statement terminator character. Pascal was widely used in scientific problem-solving applications and to a lesser extent in business applications.

SQL (Structured Query Language) was developed by developers of the DBMS as a programming language to program the databases. It has been standardized and a common set of SQL statements are implemented by all the DBMS suppliers.

ADA was developed for programming, and it was widely used in programming weapon systems. It was so named in the honor of the Lady Ada Lovelace Byron, who is credited as the first computer programmer for the work she did along with Charles Babbage. It is still the preferred programming language in programming weapon systems, even today.

These languages are generally referred to as 3GLs (3rd Generation Languages). They were also referred to as procedural languages, as they focused on the procedure for solving the problem and delivering the solution rather than on managing the CPU and the RAM. While FORTRAN, COBOL, Pascal, C, and BASIC dominated the programming scenario until the 1990s, there were many other 3GLs that were used but to a lesser extent. These are:

1. ALGOL (Algorithmic Language).
2. PL/1 (Programming Language 1).
3. LISP (List Processor).
4. PROLOG.
5. RPG—Report Program Generator.
6. There are many such other languages that were popular in their day, and perhaps they are being used somewhere.

The main features of the 3GLs were:

1. They focused on the procedure of solving the problem and their compiler handled the hardware manipulation.
2. They used the services of the OS to handle the hardware.
3. They were meant for processing bulk data that was entered offline and was made ready for processing after eliminating data-entry errors.
4. They were used in batch processing, that is, the data is transformed from the initial stage to the output after being processed sequentially by a number of programs.
5. They mostly produced paper outputs.
6. They used punched cards and mag tapes most of the time, but also used disk drives toward the end of their era.

Toward the middle of 1980s, the demand for end-user computing began rising, and facilities were built into the 3GLs to facilitate online data entry and on-screen outputs. Punched cards also faded away by that time as large capacity disk drives made inroads into computing. By the end of 1980s, end-user computing became the norm, and offline data entry gradually gave way to online data entry. Even so, 3GLs, especially, COBOL, is still being used even today with offline data entry on mainframe computers with batch processing wherever the applications involve bulk data processing. Cases like processing of traffic tickets by cops, tax collections, and collections by field agents in sectors like insurance are still being handled in this manner. But, let me hasten to clarify, even these applications are depleting with the advent of handheld computers in the form of smartphones, which facilitate all of these users to upload data on the fly from the field to the servers in the data center. While batch processing and the use of 3GLs may continue for some more time in the back-end processing, offline data entry seems to have just a short lifeline into the future.

4th Generation Languages

The demand for end-user computing increased in the organizations, as it could eliminate the paper-based transactions that needed to be digitalized later by data-entry specialists. The businesses argued that if facilities can be provided on computers, the business transactions could be performed on computers and save money spent on specialist data-entry operators. While Apple had introduced microcomputers in the late 1970s, they did not capture the imagination of organizations for use in business operations. They remained as home computers and later on as word processors. The introduction of the IBM PC (Personal Computer) in 1982 metamorphosed the thinking, and they made inroads into the organizations. They were initially used as word processors by the secretaries to the bosses. Then, Microsoft began leading the onslaught of PCs into organizations with the large set of PC makers pushing them into the organizations with the MS-DOS (Microsoft Disk Operating System). This created an environment conducive to portable software at an object code level. This environment included:

1. Thus far, the programs had to be recompiled to execute them on a different brand of computer. They were portable at "source code" level. With IBM PCs and their clones, they could be executed on a different brand of computer without recompiling, as long as it used MS-DOS OS.

2. There was a large volume of PCs on the market that used MS-DOS OS in the organizations.

3. Microsoft developed the SDK (Software Development Kit) to develop software applications for MS-DOS OS at an affordable price. Almost all popular programming languages were made available on PC. Due to the large volume of PCs, the prices of popular compilers literally crashed and made them affordable, even by home-based freelance software developers!

4. There was already a large pool of experienced software developers available in the market, and universities were offering courses on software development.

5. The IBM PC was a powerful (in those ancient days of just 35 years ago!) computer with a 16-bit processor and 128 KB RAM expandable to 1 MB!

This encouraged software developers to develop and market general purpose software as COTS (Commercial Off The Shelf) products. The programs WordStar, Lotus 1-2-3, and dBase II became very popular, and they further propelled PC sales because they gave some useful applications to organizations for practical data processing.

At around the same time, the networking of computers was coming of age, wired networks became practical, and, by combining networks and PCs, it became possible for organizations to implement organization-wide data processing with end-user computing facilities. Networked PCs became a viable alternative to costly mainframe computers.

End-user computing needed data-entry screens that were user-friendly and aesthetically appealing.

The introduction of PCs gave a fillip to graphics development, and manufacturers developed and provided excellent graphics facilities at low rates to PCs. While Apple pioneered this graphics effort, Microsoft also caught on. Apple introduced the GUI (Graphical User Interface) in the late 1980s. That can be referred to as the harbinger of the 4th Generation Languages (4GLs).

To be fair, we need to give credit to Borland and Philippe Kahn for introducing the first 4GL. Turbo Pascal was the first programming language, in my humble opinion, that introduced the concepts of a 4GL. It was developed at Borland, which was owned by Philippe Kahn. It has the following revolutionary features:

1. The existing languages needed:
   a. Write the program using a text editor.
   b. Save and close the text editor.
   c. Run the compiler and compile the program.
   d. The compiler lists all the syntax errors.
   e. Use the text editor to rectify syntax errors.
   f. Repeat steps c to e until all syntax errors are removed.
   g. Link the object file with the libraries and make the executable file.
   h. All the above are command line operations, being tedious and time-consuming.
2. Turbo Pascal came with a built-in text editor, compiler, and linker. This made it possible to:
   a. Compile the program without leaving the text editor.
   b. Rectify the syntax errors within the editor.
   c. The compiler, instead of listing all the errors at a time, showed the first error and allowed rectification, then continued compilation until another syntax error was found.
   d. Once the program compiled flawlessly, it allowed for execution of the program for testing purposes from within the editor!

Thus, Turbo Pascal offered the first version of what we now call an IDE (Integrated Development Environment). Thus began the advent of 4GLs.

The introduction of the GUI (Graphical User Interface) revolutionized the programming scenario. Different controls were developed for receiving data for different purposes, like text boxes, combo boxes, buttons, and so on. Each control had various events like click, double click, hover, change, and so on, each of which could be programmed. The key feature of the 4GL is the usage of the mouse for moving the cursor on the screen as well as for selection. While Turbo Pascal was the first 4GL, there are many others, like:

1. Power Builder
2. Visual Basic
3. C++
4. Visual C++

5. Python

6. Ruby on Rails

7. PL/SQL

8. Informix-4GL

9. Oracle Forms

There are a number of other such languages that can be used to program the GUI applications. The 4GLs have these characteristics:

1. They are focused on end-user computing.
2. They are GUI based, that is, all actions are tied to a form placed on a screen.
3. The screen is contained in a form on which multiple controls can be placed.
4. The controls have events such as click, double click, change, lost focus, got focus, and so on, each of which can be independently programmed.
5. While the keyboard can be used to navigate on screen and activate controls, the mouse is the preferred hardware device for on-screen navigation and activation or selection of controls.
6. All 4GLs have implemented some degree of fault tolerance and user assistance in programming, which can be described as parts of artificial intelligence.
7. All of them come bundled with an IDE that facilitates ease of programming and building applications. All of them provide a text editor, compiler, linker, and debugger, as well as a facility to execute the program from within the IDE. They make programming an easier and more pleasant job.
8. All of them come with a help facility to aid the programmer in language constructs, syntax, and usage at the click of a button.

While 4GLs are the result of a paradigm shift from CUI (Character User Interface) to GUI (Graphical User Interface), as well as to networked, distributed, and end-user computing, no such paradigm shift is visible on the horizon to shift gears in programming languages from 4GLs to 5GLs. But I am sure the shift will come sooner rather than later.

One paradigm shift I am visualizing right now is the possible shift from PCs and laptops as user terminals to using smartphones and tablet computers to connect to the Internet and servers at a remote location. Right now, we are developing our programs on PCs and laptops using a simulator software. In my humble opinion, we will stop using them and shift to tablets and then to smaller-sized smartphones progressively in the next two- to five-year time frame. In my humble opinion, that will be the time when we will usher in the era of 5GLs in computer programming. Will it be called computer programming, or phone programming, or smart devices programming? Only time will tell!

20

Programming Standards

Some Quotable Quotes on Coding

"If debugging is the process of removing the bugs, then programming must be the process of putting them in."

—Edgser Dijkstra

"Everything should be made as simple as possible, but not simpler."

—Albert Einstein

"Writing compact source files that makes full use of C's shortcuts operators has been a test of manhood for many C programmers. Many C hackers and even some authors of C books will tell you that you have to use all of C's features and write compact but unreadable source files. This is not true. Writing obscure, tricky programs is good for a hacker's ego but unnecessary and dangerous in serious programming projects."

—William Hunt

"I like to think the whole program through at a design level before I sit down and write any of the code... The really great programs I've written have all been the ones that I have thought about for a huge amount of time before I ever wrote them... Part of our strategy is getting the programmers to think everything through before they go to the coding phase. Writing the design document is crucial... the worst programs are the ones where the programmers doing the work don't lay a solid foundation... I really hate it when I watch some people program and I don't see them thinking."

—Bill Gates

"A great programmer loves to look at his or her own code and go through it... Greatness is the notion of always wanting to simplify, always thinking you can make it better, and really loving to look at your own code... There are some people who, once a thing works, won't go back and look at it—that's a crummy programmer."

—Bill Gates

Introduction to Standards

Why do we need standards for writing computer programs? As it is, writing computer programs involves creativity and mental work. If we place restrictions on the way we write programs, the life of a computer programmers becomes untenable! Besides, programming is a creative activity, and restrictions are the surest weapon to kill creativity!

So go the arguments put forward by the people opposing standards and guidelines for writing programs. The percentage of programmers opposing programming is significant, and it can be more than even those that support the standards!

This opposition comes from a lack of understanding of the situation and the percentage of those opposing standards dwindles as they put in more years in programming and software maintenance. As programmers begin the software maintenance work, they begin appreciating the need for programming standards. It is commonly accepted that the time taken for the development of software is insignificant compared with the time it spends in maintenance. Do you remember the Y2K problem? Initially, when programs were developed in the 1950s to the 1980s, they used only two digits to denote the year in date fields and "19" was assumed to be the century. This was done to conserve space in RAM as well as on the tape and the disk. When the century turned to 20, problems were foreseen, and huge amounts of money were spent just to rectify the two-digit year problem in the programs! Even if we assume the software developed in the 1970s, it spent more than 30 years in maintenance, and most of those programs are still in use even today and are being maintained! I am sure you can see the significance of software maintenance and the necessity to make it easier.

Let us now discuss the aspects of standards.

Standards and Guidelines

Among the multiple definitions offered by the Merriam Webster's dictionary, the one applicable for our present context says thus: "something set up by authority, custom, or general consent as a model of example." Behavioral standards are set up more by general consent and custom than by authority. The government has set up many standards for the safety of its citizens. The organizations set up standards for their products and components.

In organizations, we interpret the word "standard" as a document that is approved for use in the organization. Standards provide us these benefits:

1. They facilitate achievement of a minimum level of quality and reliability in the deliverables of the organization.
2. They facilitate uniformity between different people working on the products of the organization.
3. They facilitate quicker induction of new people into the organization. A novice can deliver the results of an expert by adhering to the standards.
4. They facilitate development of components that can be used across products on large scale and thus help us reduce the costs.
5. They facilitate easy understanding and maintenance.
6. They facilitate reliability.
7. They facilitate development of expertise and tools to improve the efficiency and productivity of persons.
8. They provide a basis and a platform from which to improve and scale greater heights in the matter of functionality, ease of use, quality, and reliability.

I am not elaborating on these aspects, as this is not a book on standards and standardization. Standards usually contain the following aspects:

1. The minimum quality that needs to be maintained by a component, material, product, or any other deliverable of organizational endeavor
2. Specifications applicable to the situation at hand
3. The methods of ensuring quality of the deliverable
4. The features and specifications for safety, security, and reliability
5. The procedural steps that need to be implemented, if applicable, in the process
6. Checklists to ensure the deliverable adheres to its specifications
7. Formats and templates to capture information uniformly
8. Any other aspects that are needed to ensure the minimum quality, reliability, safety, and security in the deliverable

Standards are developed by the standards bodies of governments (American National Standards Institute of the USA, Dueutsches Institut fur Normung of Germany, Commission de Permanente Standardisation of France, Bureau of Indian Standards of India, and so on), industry associations (National Electrical Manufacturers Association, International Telecommunications Union, International Hotels Association), and professional associations (Institute of Electrical and Electronics Engineers, American Advertising Federation of USA, Audio Engineering Society, Automobile Manufacturers Association of Japan, and so on). Then there are international bodies such as the International Organization for Standardization, International Accounting Standards Board, International Labor Organization, International Civil Aviation Organization, and so on. Of course, every professional organization develops standards for internal use besides adopting national and international standards.

In recognition of the contribution of the standards, the organizations developing standards, and the people working on the standards, a day, October 14, is dedicated every year to celebrate World Standards day.

We use the words "standards" and "guidelines" interchangeably. The difference between them is subtle. While standards are prescriptive, guidelines are suggestive. It is not an exaggeration that no activity is carried out in professional organizations without conforming to some standard. Because computer programming is relatively of recent origin, the standards are getting developed now. The IEEE released its first batch of software engineering standards in 1988 and the software development industry has been slow in adopting them. But the scenario is changing, with more and more organizations voluntarily adopting standards in their working.

Programming Standards

What do we want from developing and implementing programming standards? It is common that programmers leave the project or the organization or both in the middle of the project due to a variety of personal and official reasons. When operating in such conditions, it is imperative that we establish a set of simple coding guidelines for each of the programming languages so that the next programmer can continue the code where the first programmer left off. Without a set of coding guidelines, we would have to throw away the code written by a programmer who left the project. Here are the objectives of programming standards:

1. *To ensure ease of understanding*
2. *To ensure quality control, debugging, and software maintenance while in production.*
3. *To ensure flexibility in the functionality so that the code does not need to be changed for a slight variance in the processing logic.*
4. *To ensure efficient use of resources by the programs*: We discussed the scarcity of the resources; however, much our hardware improved and, even with the vast amounts of RAM and disk space available now compared to the 1980s, resources are always in short supply! New problems, new constraints, and new technologies gobble up RAM and disk space as fast as the makers increase them! We need to code the programs in such a way that the resources are used efficiently as well as economically. Standards help the programmers in developing code and ensuring the efficient use of computer resources.
5. *To ensure accuracy of results*: The need for precision is increasing every day. Originally, computers were used predominantly for financial and other business applications in which the precision needed was two digits after the decimal point. In the present day, most machines, including cars, airplanes and rockets are driven by computers and software, thus skyrocketing the need for better accuracy. The significant digits after the decimal point crossed two digits long ago, shooting up the need for increased precision. Standards help programmers to use the right precision for the occasion.
6. *To ensure defect prevention*: It is easier to prevent defects in programs than debugging and eliminating them. Well-developed programming standards assist us in preventing defects.
7. *To ensure reliable functioning of the software*: Most people think that reliability is not applicable to software because there are no moving parts that can suffer wear and tear. But the changes in environment, like software maintenance due to changed business or technical environment, upgrades to OS, changes in the governmental rules and regulations, changed management requirements, competition, and so on, cause our software to malfunction. With proper coding guidelines, we can increase the reliability of the software.

The benefits of establishing a set of coding guidelines are:

1. It becomes feasible to enhance the code developed earlier by a different programmer.
2. We can make use of the code written by programmers in one project by other programmers in a different project.
3. Software maintenance, which is inevitable, becomes easier for a code that is developed using a set of guidelines than a code that was developed without adhering to any guidelines.
4. We need to use multiple programmers on a software project, and coding guidelines assist us in achieving uniformity in the code produced by all the programmers.

5. Code that is written adhering to coding guidelines can be used for training new programmers in writing maintainable and reusable code.

6. Coding guidelines ensures a minimum set of quality aspects including defect prevention and efficiency of execution in the code.

Scope of These Guidelines

These coding guidelines are general in nature and can be tailored to suit the programming language we use. However, these principles would be useful for any programming language with the exception of the details. These guidelines can also be used as the overall coding guidelines, and we can have other specific coding guidelines for each of the programming languages used in our organization containing only the exceptions to this document to suit that specific programming language.

Ease of Understanding and Maintenance Guidelines

These guidelines are aimed at increasing the ease of understanding the program for the purposes of continuing the code, debugging, and maintenance. This is achieved by three guidelines:

1. Naming conventions
2. Formatting of the code
3. Inline documentation

Let us discuss each of these now.

Naming Conventions

In any computer program, we have two types of words:

1. Keywords provided by the programming language that tell the computer what to do.
2. The variable names that provide the data to be processed.

While the keywords are easy to understand, the variable names are difficult to understand for others. To know what the variable name denotes, we need to spend some time in deciphering the name. Naming conventions bring in uniformity and clarity of its meaning. To remove ambiguity and bring more clarity to variable names, we use naming conventions that define the way in which variable names are defined by the programmers. Naming conventions enable the person reading the code to distinguish between program variables, table fields, file fields, constants, flags, counters, file names, and so on quickly and implement necessary enhancements or fix defects.

Presently, most modern programming languages permit long variable names so that variable names can be named meaningfully to reflect their function. However, long variables increase the statement length and reduce programmer productivity, as it takes more time to type longer names than shorter names. We need to strike a balance between meaningfulness and brevity. The guideline is that the name must not be shorter than 5 characters or longer than 25 characters.

We use three prefix characters to denote their type and origin. It is suggested that the name be preceded by two or three prefixes. These prefixes are separated by an underscore character. In case an underscore character is not permitted by the programming language, then the first character for each of the name segments shall be a capital letter.

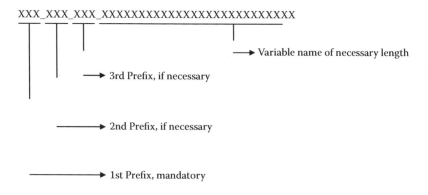

Table 20.1 shows some suggested prefixes and their meaning.

These suggested prefixes do not cover the entire spectrum, and you can add more prefixes depending on your need.

It may often be necessary to abbreviate names in programs. In such cases, it is suggested that you use three characters when abbreviating names as shown in Table 20.2.

These are by no means exhaustive, and we can add more abbreviations as necessary. Some organizations use four characters to abbreviate the names. Some use two to four characters to abbreviate. It really depends upon the organizational culture.

Here some examples of variable names in Table 20.3.

However, if necessary, if the project needs a different set of naming conventions or the customer specified a naming convention, they need to be recorded in the configuration management plan of the project and then used.

Formatting Source Code

Formatting source code improves readability and easy understanding. In the earlier days gone by when we used punched cards, we limited a program statement to one line. Not anymore! Now a program statement can span across multiple lines. That means it is possible to have multiple subordinate statements to a principal statement. Sometimes, we have a beginning statement and an end of the statement indicator keyword, such as ENDIF, or a character, such as a semicolon (;). Subordinate statements are embedded between these two principal statements. Subordinate statements can have further subordinate statements! While programming guidelines normally restrict such nesting of statement levels to 3, it is possible to have more levels of nesting occasionally. So, formatting guidelines assist us in distinguishing between the principal statements and their subordinate statements. This is essential in reviewing, debugging, and maintenance of the programs.

TABLE 20.1

Suggested Sample of Prefixes

Abbreviation	Expansion
AVR	Alphanumeric variable
CHK	Check box
CLS	Class module
CMB	Combo box
CMD	Command button
CNS	Constant
CTR	Counter
CXN	Connection
DGR	Data grid
ERM	Error message
FGR	Flexi grid
FLD	Table or data file field
FLG	Flags
LBX	List box
LVW	List view
MSG	Message box/dialog box
MTD	Method
NVR	Numeric variable
RBT	Radio buttons
RPT	Report
SBR	Subroutine/ subprogram
SCR	Screen
TBL	Table
TXT	Text box

TABLE 20.2

Suggested Sample of Abbreviation of Names

CUS	Customer
EMP	Employee
ID	Identification
LOC	Location
MAT	Material
PRJ	Project
PWD	Password
SAL	Salary
WST	Workstation
QTY	Quantity
AMT	Amount

TABLE 20.3

Sample Variable Names

Variable Name	Explanation
txt_userid	Text Box to receive user ID
cmb_prjname	Combo Box to contain project names
lvw_matcodes	List View to contain material codes
tbl_projects	Database table containing project information
fld_prj_prjid	Table field in projects table with project ID
nvr_qty_stock	Numeric Variable of quantity in stock
flg_arrayfull	Flag to check if the array is full
ctr_items	Counter to count number of items

Distinguishing the Principal Statements from the Subordinate Statements

Whenever there are nested control flows (if, while, case, for, etc.), we separate the subordinate statements by moving the left margin of the subordinate statement by one tab-character length. The left margin of each subordinate level would be moved rightwards by one tab. This is shown in this example:

This is the principal statement.
Part of the principal statements.
    This is the first level of subordinate statements.
    Any first level subordinate statements.
        This is the second level of subordinate statements.
        Second level subordinate statements.
        This is the closing statement for second level subordinate statements.
    Any more first level subordinate statements.
    This is the closing of first level of subordinate statements.
Any more principal statements.
This is the closing principal statement.

This would help us in correctly identifying the program logic, the order of execution, and the control flow of the program.

Limiting the Length of the Line Such That It Becomes Easily Readable

Modern programming languages permit longer line lengths, up to 255 characters per line. Similarly, modern screens also permit longer lines. The length of a programming line must be limited to the length of the line permitted by the screen. There should not be any need for horizontal scrolling to read the program. If lines longer than the screen width are required, they may be broken down to multiple lines using the statement continue convention permitted by the programming language. However, the continuation lines must be treated as the subordinate statements described in the earlier section, and their left margin ought to be offset by one tab-character length.

Separating Segments of the Code

Separating the code segments using separators makes it easy to understand the functionality of the program. The separator shall be a commented statement. Additionally, including a blank line would make the program more readable. Here are some examples assuming "/*" and "*/" as the first and the last commenting character sets for commenting statements:

```
/* Begin the case statement to select the appropriate category of the
ticket */
/* End the case statement to select the appropriate category of the
ticket */
/* Begin the loop for reading all the records in the table */
/* End the loop for reading all the records in the table */
```

Inline Documentation and Commenting

We need to explain the logic for each code segment using the commenting feature of the programming language so that the software maintenance people will not have any problems in understanding the logic of the program. Every program shall have the following inline documentation in all their programs:

1. Each program shall have a header. This header will contain the following:
   a. The name of the program.
   b. The organization name that developed this program.
   c. The date of beginning the initial coding of this program.
   d. The functionality achieved by the program, in brief.
   e. References of calling programs and the programs called from this program, if any.
   f. The revision history of the program containing the following information for each modification:
      i.   The date of modification.
      ii. The name of the programmer who made the modification.
      iii. Description of the modification.
2. Each control statement would have an explanation of the purpose of the control statement and the expected results.
3. Each of the loops, especially those that are used for reading all the records from tables, would have explanation at the beginning and at the end of the loop.
4. Each subroutine/subprogram would have an explanation of the purpose of the subroutine/subprogram and an explanation of the parameters required by this subroutine/subprogram. It would also explain the expected parameters that are to be received as well as the values returned by it, if any.

Commenting Style

1. As much as possible, comment and code should not be mixed in the same line. The comment should precede the concerned statement.

2. Keep the length of the commenting line to the length permitted by the screen. There should be no necessity to scroll the screen horizontally for reading the comments.

3. As much as possible, do not spread the comment across multiple lines. Each commenting line should be self-contained. If more than one comment line is required, the "#" character shall be used at the end of the previous line to indicate that the comment is continued on to the next line. Here is a two-line commenting statement for example:

```
/* Copy right = Company Name #*/
/*New York */
```

Program Header Example

```
/* Name of the organization = Chemuturi Consultants   */
/* Program ID / Name = Item issue screen              */
/* Original Author = Chemuturi Consultants            */
/* Original Creation Date = 1 Nov 2017                */
/* Parameter List =                                   */
/* Program explanation                                */
/* IPR rights belong to = Chemuturi Consultants       */
```

Example of Revision History Documentation

```
/* Revision History */
/*********************************************************************/
/* Programmer Date Description of modification */
/* ***************************************************************** */
/* ***************************************************************** */
```

Example of Loop Control Documentation

```
/* Purpose = */
/* Entry Condition = */
/* Exit Condition = */
```

Special Inline Documentation

Whenever, there is complexity in the code, especially when writing long computational statements or a procedure for complex decision-making or mathematical processing, appropriate inline documentation may be included.

Efficient Use of Resource Guidelines

We use these guidelines to ensure that the system resources are used efficiently and economically.

Declaration Statements

We use these statements to declare variables and other objects as necessary. Each declared variable takes up space in the RAM while under execution, so we should declare variables carefully to conserve RAM. The following are the guidelines for this type of statement.

1. Declare only one type of variable or object for each statement.
2. Code all declaration statements at the beginning of the program.
3. It is not preferable to mix declarations. Code declarations in groups, when used, they ought to be in the order given in the following:
   a. All numeric variables of integer type
   b. All single-precision floating-point numeric variables
   c. All double-precision floating-point variables
   d. All date-type variables
   e. All alphanumeric variables
   f. All numeric arrays
   g. All alphanumeric arrays
   h. All database objects including connections and tables

Defect Prevention Guidelines

The way we write programs can leave loopholes for defects to creep in. Here are some guidelines that would plug loopholes and prevent defects from creeping in.

Control Statements

Control statements are one source of defect injection. When we use control structures without diligent care, the program execution may not go through the path assumed by us and lead to erroneous results or failures. The following guidelines help in ensuring that control structures are properly coded to prevent defects.

1. Using the right control structure would go a long way in preventing the defects. Here are some guidelines for selecting the right control structure.
   a. Use "case" structure when multiple courses of action are available based on the result of one condition.
   b. Use an "if" control statement when a set of statements have to be executed only once, depending on one or more conditions (logical expressions).
   c. Use a "for" loop when the maximum number of iterations for the loop is finite and known beforehand.
   d. Use a "while" loop when the maximum number of iterations for the loop is not known before hand and is dependent on a condition. There are two kinds of "while" loops, with one of them checking the condition at the start of the loop

and the other at the end of the loop. It is preferable to use the loop that checks the condition at the start of the loop.

    e. Avoid using the "goto" structure as much as possible. The reasons are:

        i. It leads to free-fall execution of the program, and it is difficult to predict the course of execution.

        ii. If it results in closing the program, we may not be able to control the cleanup activities before smoothly closing the program.

    f. Use of "goto" structure is permitted only in the case of error-trapping statements.

    g. Ensure that a program has only one entry and one exit. It is preferable that the same code segment has the entry point and also the exit point. The program execution control is exercised from this segment to other segments and finally exits from this segment. Even the error-trapping statements need to pass the execution control to this exit point when necessary or if closing the program is the only option available.

2. When using "if" statements, these precautions are necessary:

    a. Always code the "else" part of the statement. While we may not be able to see the possibility, the conditions in the field are always beyond our comprehension and the unthinkable always happens. Therefore, coding the "else" part of the "if" statement helps in the prevention of defects.

    b. It may become necessary to nest several "if" statements in our programs, that is, inserting another "if" statement within an "if" statement. In such cases, limit the level of subordinate "if" statements to a maximum of three levels. This means one main "if" statement and two subordinate "if" statements, totaling a maximum of three "if" statements together in one nest. If it becomes necessary to have more levels, use "case" structure, break up the program, or take another look at the program design.

3. When using a "while" loop, ensure that the condition has a probability of becoming true (or false, as the case may be) so that there is an exit point for the loop. It is very easy to enter into an infinite loop by using the "while" structure. This loop is the one that is used to read the records from a file or table until the EOF (End of File) condition is reached, and it is often forgotten to move the record pointer forward in each iteration, leaving the loop processing only one record infinitely.

4. When nesting the "while" loops, again, limit the nesting to a maximum of three. It is preferable to call a subroutine for each nesting of "while" loop rather than code all the statements together, as this loop normally uses a number of statements as compared to other loops that use a far less number of statements.

5. When using the "case" structure, always code the "default" option (that is when none of the values mentioned in each of the cases is valid). This prevents free-falling of the program execution.

Loops

Loops are often a source of defects. We often come across three possible defects in loops:

1. *Absence of an exit condition inside the loop*: We forget to include a statement inside the loop that makes the condition true so that the loop can be exited. This often happens when we try to access a resource that is locked by another program, switched off, or faulty. In such cases, it is a common practice to use a timer or some such mechanism to time out the checking and exit the loop. But often, programmers forget this statement, and the loop becomes infinite. They do that because the device is expected to be there waiting for the program, but sometimes the device will not respond due to various reasons. We need to make certain that we have included a loop exit statement inside a loop, especially when we are looking for a device.

2. *Not incrementing the counter*: In finite loops based on counting, we forget to increment the counter inside the loop. In a For...Next loop, the counter is in the statement and is incremented automatically, but in While... loops, we need to declare a counter and then explicitly increment/decrement the counter using a program statement. Sometimes we forget this statement and it becomes an infinite loop. We must ensure that a counter-incrementing statement is included inside the loop.

3. *Reading empty table or file*: As most of the programs use data from flat files or database tables, we use loops to read the records and process them one by one. Usually we use the While... loop to perform this function. We expect that there are some records in the file or table, especially during the first use when the table or the file is empty and the read statement fails and causes a fault. Therefore, every time we open a file or table for reading, we need to include a statement that checks the table or file for the EOF (End of File) condition. We need to include other statements along with this statement to tell the computer what to do in case the file or table is empty. This will prevent the fault from developing due to empty table or file.

One more possibility of developing faults in the loops is faulty initialization of variables. For the loop to terminate, we need to either have a counter reach its final value or for a condition to become true. For both the cases, we may use a variable, which needs to be incremented. If the loop does not increment the counting variable automatically, we need to write statements incrementing the counter inside the loop. A common error in loops reading records from a table of a file is not to increment the record counter, thereby reading only the first record always and entering an infinite loop. We need to always ensure that the concerned counter is incremented inside the loop.

One other possibility of developing faults in loop statements is using a loop inside a loop. It is better not to use a loop inside another loop. Calling another loop would be risky because that loop may become an infinite loop, or it can make the parent loop (the calling loop) to exit by making its loop condition to true. But I agree that nesting of loops often becomes essential. We should be very careful to see:

1. The child loop shall not become infinite.
2. The child loop shall not render the parent loop to exit by changing the loop exit condition to true.

Computational Statements

Computational statements, just to remind you, are used to resolve mathematical formulas. Computational statements, especially the long ones, are likely to inject defects into the program execution. One of the reasons is the order of computer processing of the arithmetical operators is difficult to perceive. Second, the results from the computation are also difficult to predict. The following guidelines help in preventing defects arising out of improper coding of computational statements:

1. One of the danger zones is the denominator in a division arithmetic statement becoming zero. This will lead to program failure. Worse still is both numerator and denominator becoming zero. When there is a division operator in the equation, check if the denominator is zero before performing the division. It is better to perform division on a standalone statement.
2. Precision problems can occur when we are multiplying two or more quantities, and we need to ensure that the variable receiving the result is large enough to hold the result. This issue can be prevented at the design stage by finding out the largest possible values that would be used in multiplication and providing variables of appropriate size both in the program as well as in the database for storage.
3. Writing long computational statements is conducive to injecting defects in the program. Use the following guidelines:
    a. Limit nesting of the parenthesis to a maximum of three—three opening parentheses and three corresponding closing parentheses.
    b. Limit each computational statement to be visible without the necessity to scroll the screen horizontally. Instead of coding a long computational statement, break it down into multiple computational statements.
    c. Code arithmetic division in a separate statement as much as possible, and write a preceding statement checking if the denominator is zero.
4. When mixing the arithmetic operators of addition, subtraction, multiplication, and division, do not assume the default operator precedence of the computer. Use parentheses very liberally. Placing the addition and subtraction operations inside a parenthesis if mixing of operators is especially imperative.
5. Division and sometimes multiplication operations using floating-point variables cause rounding problems. In such cases, two possibilities exist for preventing defects:
    a. Always round off to one or two more digits of precision than required and finally, before either presenting the result or storing it, round off to the required precision.

b. Carry out the operation using whole numbers and convert them to decimals by dividing by 100 (or 10 or 1000 or 10,000 etc.) for presentation or storage purposes as much as possible.

6. Duplication of routines is another common cause for injecting errors. When the same operations are to be performed in multiple programs, some programmers duplicate the routine. It is always better to keep one routine and use it in all places by passing appropriate parameters to it. This protects the integrity of processing and prevents defects from creeping in.

7. As much as possible, do not use table/data file fields in computational statements, especially to receive results of computation. Always copy the value of data file/table field into a variable and then use it in computations. Similarly, receive the value of computation into a variable and then move it to data file/table file field just before writing it.

8. When rounding off the value, code the statement on a separate line just for rounding off the variable. Do not use the rounding function in combination with a computational statement.

Efficiency Guidelines

Efficiency guidelines help us in ensuring the efficiency of execution as well as using the resources of the computer economically, especially the RAM. The following guidelines would help:

1. Do not declare any variable or constant without any purpose. It is common practice among programmers to declare a number of variables with a view that they may be necessary in the program. Avoid the temptation to declare too many variables; even though the stringency on resource usage is now a thing of the past, occupying too much RAM is likely to slow down the computer.

2. As much as possible, declare variables as local to the program and use parameters to pass values to other programs or subprograms. When we declare variables as local variables, their RAM would be released on exit from the routine. If we declare variables as global variables, they would hold on to the RAM until we stop execution of the entire set of programs.

3. Open the files (or database tables) only when required, that is, just before the file operation statements begin, and close them as soon as the file operation statements end. Opening a file or database table occupies a chunk of memory and it takes CPU time, too, to keep checking the file status/condition. It may also prevent other concurrent users from accessing the files/tables.

4. Keep a limit on the number of objects that can be kept open concurrently as they take up large chunks of RAM, which slows down the program execution.

5. Also, do not pack too many controls onto one screen. Instead, divide the screen into multiple screens/tabs. This would reduce the burden on the RAM usage.

6. As much as possible, do not print directly from the program. Program-controlled printing is not very efficient. When printing directly from the program, we need to control the printer and trap its errors such as out of paper, out of ink, and out of power. Otherwise, the printer may cause program failure. Unless it is receipt- or ticket-printing needed across the counter, create print files so that they can be printed using printing utilities of the operating system, which are much more efficient in printer control and printing.

Effectiveness Guidelines

These guidelines help us in using the software effectively. The following guidelines helps us in doing so.

1. In case of bulk data-processing applications, appropriate control statistics, such as the number of records processed, records included in the report, control totals, etc., are generated on suitable media and delivered to the user.

2. When coding screens for user input/query, the background/foreground contrast must be significant to ensure easy readability.

3. For input screens, unless the client insists, the cursor shall not move from field to field automatically, and when end-of-field is reached, an audio signal may be generated to alert the user about the end-of-field condition.

4. Try to use more statements than a fewer number of statements so that the program would be easily understood during software maintenance. Please note that whether you code a complex single line or simple multiple lines, both will translate ultimately to machine instructions, and the number of machine instructions in both cases would almost be the same.

These guidelines are prepared as a starting point for you to develop your own coding guidelines that are best suited for your organization. You may use these guidelines as they are now, or add to them, modify them, or remove some of them at your free will. What I suggest is that you have guidelines for code consistency, defect prevention, and efficiency and effectiveness aspects.

21

Personal Software Process

Introduction

Every employee wants to work in a methodical manner without having to go back and forth making mistakes and correcting them. The employee also wants to produce deliverables that have zero defects so that no other person can point out a mistake in the deliverable. But the reality is different. We commit mistakes, we do go back and forth making and correcting our mistakes. Others, especially the quality control persons, do point out our mistakes. To err is be human, and we are human beings and commit mistakes even against our will. We do not have all the time in the world to do the best possible job. We do have deadlines to meet and deliveries to make. There are other pressures at the workplace, like competing for awards.

There are three aspects to working:

1. Complete the deliverable.
2. Complete the deliverable on time.
3. Complete the deliverable with no defects.

We need to meet all the earlier three aspects while working in organizations. Then there are three levels of acceptable performance:

1. *Penalty-avoidance level of performance*: This level of performance delivers with just a wee bit more than acceptable delays and defects, but not so off the mark that penalty becomes due. The bosses recognize the person as an underperformer and may place the person on a performance improvement plan.
2. *Normal performance*: This performance delivers acceptable level of defects and acceptable level of delays of delivery. That is, there are defects in the work product, but they are within the organizational average. The delay in delivery is not much and, in fact, it is considered as on-time delivery.
3. *Award level of performance*: This performance always delivers the work product ahead of the deadlines and with lesser defects than the organizational defect level. It is above the organizational average performance.

In organizations, most employees are in the first two levels, namely, the penalty-avoidance level and normal performance level of performance. There would be but only a few that will be performing at award level. There are various reasons for the performance being low. These could range from a lack of personal ambition, lack of self-motivation, and lack of knowledge about how to improve performance. My experience shows me that:

1. Every employee comes to work to "work" and effect great deliveries.
2. Every employee puts in his/her best effort.
3. Every employee yearns for recognition.

So, the organizational environment, lack of trained supervision, absence of a well-designed and administered reward system, and the absence of a grievances resolution system effects the motivation levels of the employees. Motivating the employees toward a better performance is a combined responsibility of the organization and the individual. This book is not the right forum to discuss the organizational role in employee motivation, but it is certainly the place to discuss the motivation of the individual. I am going to do this now.

Why does an individual underperform? It is due to two factors. One is the set of organizational factors, and the other is the set of individual factors. I believe these are the individual factors responsible for underperformance:

1. Lack of proper training.
2. Lack of knowledge on how to improve performance.
3. Lack of capacity to do better.

Every individual has limitations, both physiological and psychological, on their capacity to do work. These are genetic factors that the individual has since birth. It is extremely difficult to correct these congenital defects, but the individual with grit and commitment can overcome these. Lack of proper training can be alleviated with training, either on the job or in a classroom. This is very easy to correct and bridge the gap. What usually happens is that all individuals perform equally at the time they join the workforce, but some gallop ahead while others lag behind. This happens because those that galloped ahead knew how to improve their performance either by structured instruction, parental guidance, or by sheer intuition. For those of you that do not know how to improve your performance and also how to highlight that improved performance, I am discussing the path in this chapter.

Personal Software Process

A process is a series of steps to achieve a preset goal. It may consist of procedures, standards, guidelines, formats, templates, and checklists. Every professional software development organization has a well-defined and continuously improving process which is adhered to for all work that is carried out in the organization. SEI (Software Engineering Institute) of CMU (Carnegie Melon University) developed process guidelines and arranges for the appraisal of organizations for adherence to those guidelines. SEI also developed a set of process guidelines for individual programmers and labeled them as the Personal Software Process. This discussion is an adaptation of that process. I have developed my own guidelines for individual programmers to adopt and improve their performance.

There are two aspects to any work, and they are productivity and quality. Productivity is the rate of achievement, and quality is the presence or absence of defects. Productivity is measured in the amount of work performed per hour or per day for individuals. Quality is measured as the number of defects per million opportunities. For example, if you wrote a program with, let us assume, 1000 lines of code and committed 3 errors, then your quality is 3 defects per thousand lines of code, or 0.3 defects per 100 lines of code or 3000 defects per a million lines of code! The world of programming strives to achieve the level of 3 defects per a million lines of code. Of course, the target of 3 defects per a million lines of code is for the delivered lines of code that were subjected to rigorous quality-control activities. This measure is also called the defect density.

So, basically, your programming work is measured by the number of lines of code developed per day and the defect density in those lines of code. Therefore, to improve your performance, you need to develop a higher number of lines of code (increasing productivity) and reduce the number of defects in those lines of code (decrease the defect density). How do you do this? Let us discuss.

Productivity

We receive a salary as compensation for carrying out the work assigned to us in the organization. The value of our work should not only earn the salary paid to us but also some more money to cover the overhead and result in some profit for the organization to be passed on to the entrepreneurs who invested in our company. Initially, we may not be able to fulfill this obligation, but as time passes, we need to improve our expertise as well as our productivity and fulfill this obligation. To be able to achieve this target, we need to track our productivity on a regular basis and see the trend in our productivity.

In order to derive our productivity, we need two items of data, namely, the size of the programs we developed, and the time taken for developing them. Then, productivity can be derived using the formula:

Productivity = Size of programs developed/The time spent in developing those programs

We need some explanation to understand the earlier formula.

How do we measure the size of the programs? There is difference of opinion about which unit of measure is apt for measuring the size of the programs. Function points have a majority backing as the unit of measuring the size of software; it is mostly used at the software-product level rather than at the individual program level. At the program level, the LOC (Lines of Code), in my humble opinion, is the best unit of measure. On the LOC, there is again some difference of opinion on these aspects:

1. Should we include commenting (or in-line documentation) in the count of the number of LOC?

2. Should we take a statement as one line or would each physical line be counted as one LOC?

3. There are some very short lines of code, such as declaration of variables, and some very long lines, such as arithmetic expressions. Would both be counted as one line each, or should we assign some weight to the LOC?

4. Many languages allow for the declaration of multiple variables, including their initialization, in one line itself. We can also declare just one variable in one line and initialize it in another line. Should we declare each variable as two lines if we declare and initialize multiple variables in one line?

Here is my take on these points.

1. I suggest that the commenting lines need to be counted as the LOC because they have a purpose, even if they are not processed by the computer. They assist the quality-control persons in understanding the program, and also the programmers in the future for the easy maintenance of code. You also spend time thinking and forming brief sentences to convey the meaning accurately and precisely. They consume your time and they are mandatory in any professional organization.

2. Again, I suggest that you take the physical line as one LOC. If you take a statement, it may contain blocks of statements within it. For example, in loops, subloops can exist. Similarly, in control statements, other control statements and even loops may be there. So, it would be accurate to take a physical line as one LOC. The situation is the same for all programmers.

3. In the case of short statements and long statements, I would suggest showing no difference for counting the LOC. If we begin assigning weights to lines based on length, the data collection would become tedious and time-consuming. The counting of the LOC should not to be so rigorous that it becomes more time-consuming that writing of the lines themselves! Obviously, on the whole, the short lines and the long lines would average out. The other aspect is that all the lines in a program would never be of the same length.

4. Yes, we can declare one variable per line or multiple variables in the same line. We can also initialize the variable during declaration in the same line or on a separate line. The choice is ours. But in this aspect, the organizational coding standards would specify how to declare and initialize variables. If they specify one variable per line we code accordingly, and if they specify multiple variables per line, we code in that manner. Since our coding practices are in adherence to the organizational coding standards, we count just the physical lines irrespective of the fact that multiple variables are declared and initialized in the same, single line.

I am sure that a professional programmer would never include blank or unnecessary statements in the program just to boost their productivity. Also, a professional organization would certainly subject the code to peer review, and if there are any unnecessary lines in the program, they would be pointed out and corrected. There is an advantage in counting physical lines because you can easily develop a small utility to count physical lines of code. All you need to count is the number of CR characters. Better still, most IDEs give the line count and we do not have to develop a special utility for that purpose.

When we come to time spent on developing the programs, the questions arise as to:

1. Should we include the time spent in reading the specifications or the design document in the time taken for developing the program?

2. Sometimes, we just need to take some time thinking about the problem or how to achieve the functionality. Should we include that time also?

3. How about the time we spent in answering queries from the project leader or bosses about the progress of the work?

4. How about the time we spend on personal needs like drinking water at the water fountain or drinking coffee?

5. What about the time that is wasted on various workplace disturbances?

I suggest that you include all those times because you are getting paid for those times, too. I would say that the only times to be excluded are those times that are wasted as a result of organization-wide disturbances when no one was able to work. Except that, you include all the time you spent from the time you begin working on the program until you release it to the quality-control activities.

Now, productivity needs to be computed at two stages. The first one is the initial coding. We need to take the time and the size when we completed the coding and submitted it to the quality-control activities. This is referred to as the initial coding productivity. The quality-control activities uncover defects, and we may insert some more code in fixing those defects, so we need to compute the second productivity metric after the completion of the quality-control activities. This is referred to as the final productivity metric in the industry.

Quality

Quality is not easily amenable to measurement. A customer expects no defects in the deliverable, but experience shows that zero-defect delivery is rather a goal than a reality. So, the quality is measured in the delivered defects per unit of delivered product or service. Now, the final delivery is affected only after the deliverable is subjected to quality-control activities and rectifying the defects uncovered during the quality control activities. Even then, some defects still linger on inside of the deliverable. Quality assurance of the deliverables is a large subject in itself and is out of scope for this book. For us, the computer programmers, quality is important and is our responsibility because the quality-control activities only "uncover" the defects but never correct them! We have to build the deliverable without defects in the first place and then rectify the defects when pointed out by the quality-control activities. Therefore, we need to measure our quality and then continuously improve it.

For our purposes, let us define the quality of the deliverable as the number of defects uncovered in our work per every 100 lines of code that was written by us. This measure is referred to as the "defect density" in the industry. The formula for computing the defect density is:

$$\text{Defect density} = (\text{Number of defects} \div \text{Total LOC}) \times 100$$

Number of defects: It is the number of defects that are uncovered after you submitted the program to quality control for peer review and testing. The defects might be uncovered in review, unit testing, integration testing, system testing, acceptance testing, and any other testing that our deliverable was subjected to. We would certainly get the data for the quality-control activities that are immediately conducted on our deliverable, but we may not get the data for later testing activities. We ought to collect as much data as is available and compute our defect density.

Just as in productivity, we need to compute the initial quality metric when the first set of quality-control activities on our program are completed and the final quality metric when all the quality-control activities are completed.

Schedule

Schedule, or the date of delivery, is another important facet of our working. Whenever work is allocated to us, invariably it is accompanied by a date by which it needs to be completed and delivered. This delivery date is assigned, and it is expected of us to complete the work by that date and submit the artifact to the quality-control people. The quality control takes its own time and is subject to another schedule, so we need to restrict our schedule to delivering our code to the quality-control activities. An important aspect of this assumption is that we would not deliver defective code intentionally, and the code would only have the defects that are on par with the defect density standard of the organization. Now, the schedule metric is computed using the formula:

$$\text{Schedule metric} = \text{Number of days allocated for delivery}$$

$$\div \text{Number of days in which it is delivered}$$

There is a question here—the date is the significant one, not the number of days taken for it! Right? See, when a supervisor or project leader allocates work, the beginning date and the ending date are usually specified by him/her. More often than not, we would be able to begin work on the specified date, but sometimes, due to vagaries of the organizational environment, we may not be able to begin work on that specific artifact on the scheduled day. If the beginning schedule is slipped, the completion schedule would also slip. If we take only the date as sacrosanct, our schedule metric would be in the red most of the times. That is why the schedule metric is always computed using the number of days spent on the artifact.

Data Collection

To be able to compute the above metrics, we need data. We need to collect that data, as we are the source of the data and we need to collect meticulously. This data will help us to monitor our progress toward excellence. I suggest using a spreadsheet like MS-Excel to collect the data. I suggest using this format shown in Table 21.1 in which to collect the data.

I suggest that you maintain this spreadsheet in ascending chronological order. That way, we can easily plot a trend graph. It would be better if you began a new spreadsheet every year so that you can compare year-on-year data and see the trend.

Figure 21.1 shows an illustrative trend graph for the initial and final productivity metrics for assumed data. This graph is given only for illustrative purposes, and the data shown therein should not be taken as real-life data.

TABLE 21.1

Work Register or Data Collection Format for Computing the Metrics

Name of the Artifact	Scheduled Start Date	Scheduled End Date	Actual Start Date	Actual End Date	Schedule Metric	Initial LOC	Initial Effort in PH	Final LOC	Final Effort in PH	Initial Productivity Metric	Final Productivity Metric	Initial Defect Density	Final Defect Density

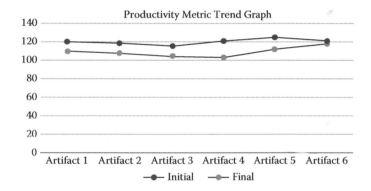

FIGURE 21.1
Trend graph for the initial and final productivity metrics.

Using similar methodology, we need to plot the graphs for defect density and schedule metrics. We can compute periodical metrics for these three categories and then compare the improvements in our performance. I suggest that you compute the metrics once every quarter, as once every month is too short and we may not have completed many programs to draw inferences based on the averages. Six months may be too long, as we lose opportunities for improvement based on factual data. You may select the periodicity based on your environment and unique situation.

You may use these formulas for computing these metrics periodically:

$$\text{Productivity metric} = \text{Sum of LOC coded during the period} \div \text{Total time spent on coding in person hours}$$

$$\text{Defect density} = \left(\frac{\text{Total number of defects in the period} \div}{\text{Total LOC coded during the period}} \right) \times 100$$

$$\text{Schedule metric} = \text{Total number of days allocated for delivery} \div \text{Total number of actual days taken for the delivery}$$

The data needed for computing these metrics is available in the spreadsheet suggested in the previous section if you maintain it regularly. You ought to take these precautions for this process to be effective:

1. Be totally honest while collecting data. After all, you are computing these metrics to improve yourself and your performance. Unless the data is accurate, the results and the inferences cannot be accurate.

2. Compute these metrics regularly in the periodicity selected by you. If you are lax in generating these metrics, you lose valuable feedback on your performance.

3. Set improvement goals realistically. Based on your capacity, set the improvement goals. For example, there is no point in setting a 100% improvement goal for the next period as it is not achievable. Similarly, setting a 5% improvement goal is also counterproductive, as that can be the margin of error in the measurement.

4. Normalize the data before you compute and contrast your metrics. If you have worked on a platform that is totally new to you, it is better to take that data out of consideration as it would drag down your metrics. Similarly, do not consider data that includes some known hindrance that slowed down your performance. Perhaps there was a hold from the customer, and because of this, you were forced to be idle for a day or two. This is not your fault.

Now, how do we interpret these metrics? Here is the explanation:

1. Productivity metrics ought to be increasing. That is, the LOC per person hour or person day must be increasing. If you achieved 100 LOC per person day in one period, it must increase to 101 LOC or more during the next measurement. If it goes down below 100 LOC in this example, then your performance is degrading!

2. Defect density must be diminishing. That is, if you had 2 defects per 100 LOC in the first period, they must diminish to 1 or lower in the next measurement period. If they go above 2, then your performance degraded. When we measure defect density per each 100 LOC, it may be in fractional numbers. We should show improvement, at least in fractions.

3. The schedule metric needs to be as close to 1 as possible. If it is more than 1, it means that we delivered ahead of schedule. If it is less than 1, it means that we missed the delivery date and delivered late.

We perform activities other than writing software code during our working time. We also conduct peer review on the code developed by our colleagues as well as test the code written by our colleagues. It is better to maintain separate spreadsheets for different activities and compute these three metrics suggested in the previous section.

Methodology

The effort needed to perform a task depends on the methodology used to perform the task. If we code a program taking a design document as the basis, we achieve one kind of results, and if we code a program based on the explanation given by the designer or project leader, we get a totally different set of results. This is true especially for productivity metrics and schedule metrics. Defect metrics are not affected by the methodology used. Then again, developing a program using coding standards takes a different amount effort than when we code a program without adhering to any coding standards. How do we handle these aberrations?

I suggest that you maintain different spreadsheets for different methodologies of working. Ideally, we ought to follow one methodology in software development, and most professionally managed organizations do follow one documented and continuously improved methodology across all the projects in the organization, but the specific organization you work for may not be one such organization. If your organization follows a documented software development process for all projects, then you can maintain one spreadsheet. But if your organization follows different methodologies for different projects based on customer preference, you better maintain different spreadsheets. Otherwise, you are likely to get erroneous values and you really cannot understand if you are improving or not.

Is it necessary to document your own development methodology? I would say yes, but...

It is better to document the methodology you adopt in performing your work. It need not be an elaborate document. You simply enumerate the steps you go through in performing your work. Here are some guidelines for different activities for you to consider but have your own methodology based on your unique situation.

Coding Methodology

Here are the guidelines for the coding process:

1. Study the design document if there is one, or spend some time in understanding the explanation given by the project leader.
2. Contemplate on the algorithms that need to be used and finalize the algorithms.
3. Open the IDE and set up the programming environment.
4. Load the form if made available by the graphics designers or lay out the form on the screen.
5. Begin programming:
    a. First, code the form load event.
    b. Then, code the events, beginning with the control at the top left-hand corner, move to the right, and then move downwards if the right end is reached.
    c. Code the "save/update" button at the end.
6. When coding the events:
    a. First, declare the variables.
    b. Then, code the algorithms.
    c. For each event, check if there is a piece of code available for reuse and use it wherever possible.
    d. When event coding is completed, remove unused variables and trash code, if any.
7. Review the code to ensure that the results expected of the code would be delivered effectively and efficiently and in conformance with the organizational coding guidelines.

8. Conduct white-box unit testing of the entire code and fix any errors uncovered.

9. Submit the code to the project leader or quality control for carrying out quality-control activities.

Of course, you can make it more elaborate or abridge the earlier guidelines to suit your unique situation. You can add more guidelines or remove some of the guidelines depending upon your organizational environment.

Peer Review Methodology

I suggest these guidelines for peer review conducted by you:

1. First, study the design document or discuss the functionality with the author of that program and learn its functionality.

2. Open the artifact in its IDE.

3. Scroll the entire program and ensure that the formatting of the code adheres to organizational coding guidelines. Record the mistakes in formatting if any in the prescribed review report format.

4. Review the form load event program and note down any errors.

5. Review the code of all controls beginning with the control on the left-hand side top corner and move toward the right and downwards progressively.

6. Ensure that all necessary events of each control are coded.

7. While reviewing the code of all the controls including the form, ensure that the algorithm used is appropriate for the scenario at hand. Note down suggestions for improvement, if any.

8. While reviewing the code, ensure compliance to organizational coding guidelines for defect prevention, efficiency, effectiveness, and the accuracy of the results.

9. While reviewing the code, ensure that no trash code (the code that should not be there) is present.

10. While reviewing the code, ensure that no malicious code is present in the code anywhere.

11. Prepare the review report.

12. Record the opportunities for improvement, if you noticed any, and include them in the review report.

13. Verify it for accuracy of the defects pointed out.

14. Hand over the review report to the project leader or the author of the artifact.

You can improve the earlier guidelines to suit your unique environment by adding or dropping some of the guidelines as necessary.

Testing Methodology

I suggest these guidelines for conducting your testing work:

1. Study the design document for the artifact, if there is one. Otherwise, discuss the functionality of the artifact given to you for testing with the author of the artifact.

2. Obtain and study the testing guidelines for the type of testing to be carried out. The type of testing could be unit testing, integration testing, system testing, negative testing, stress testing, or any other type of testing to understand what is expected of you.

3. If the test environment is already set up, study the testing environment so that you can conduct the testing effectively and efficiently. If the test environment is not in existence, plan and set it up in consultation with your project leader or the author of the artifact.

4. Obtain the test plan and test cases, if they are made ready, and study how you need to go about testing the artifact. If there is no test plan or test cases, you need to prepare one. Of course, the test plan and test cases need not be elaborate, but they need to be comprehensive so that all aspects of the artifact can be thoroughly tested.

5. Keep the test report format, either in soft copy or hard copy, ready to record the test results. Enter all the test cases in it, along with the expected results for each test case.

6. Conduct the testing and record all the instances where the actual results deviate from the expected results.

7. Once testing is completed, review the results of each test case, contrasting the expected and actual results to decide if the test case passed or failed. That is, the actual result of your testing is the same as the expected result. If the testing did not pass, then denote the test case as failed in the manner required by the testing report format.

8. Review the report and then sign it off.

9. Hand over the test report to the project leader or the author of the artifact as required and take up new task.

Of course, you may improve the earlier guidelines as necessary to suit your unique environment. You may add, delete, or modify the guidelines as necessary.

Housekeeping

You need to do some housekeeping in order to deliver excellent results continuously. This will help you to learn about your own performance. All of us perceive that our performance is top notch, but until we compare our performance with that of others, we will never know where we stand in comparison with our peers. Everybody is unique, but when it comes to on-the-job performance, our performance needs to be at par with that of our colleagues who have similar qualifications and experience, drawing a comparable salary. You need to perform these ancillary activities in addition to the main activities of carrying out your main software engineering activities. Here they are:

1. *Recording details of work*: Whenever some work is allocated to you, the first thing you need to do is make an entry in the work register described in the previous section of this chapter. It should not take more than five minutes of your time. Then, when you complete the assigned work, that is, when you are about to return the deliverable, complete the remaining entries in the work register for the completed task. This will enable you to conduct an analysis of your performance whenever you wish to. Do this diligently.

2. *Periodic measurement and analysis*: Most software development organizations require you to submit a weekly performance report. Some organizations may ask you for a monthly performance report. Even if your organization does not require you to submit a periodic performance report, set a specific interval to carry out analysis of your performance based on the data collected in your work register. Let the interval be the same as that specified by your organization. If there is no such interval, my suggestion is to set it up as once in a month. Compute all the metrics for each individual artifact and then cumulative metrics for the month. Then make a comparison with the metrics of the previous period and see if there is an improvement. If you find that your performance improved, it is good. If it deteriorated, then analyze your situation to understand what caused the downfall. Then make an improvement in the next period to bring about improvement in your performance.

3. *Setting goals*: Whenever you measure and analyze your performance, you need to set goals for your improvement. Do not set the goals too low—that is, up to 5%. The variance could be due to measurement error. Do not set up the goals too ambitiously—that is, a 25% improvement, which is not feasible in short term. Set up the goals above 5% but below 20% for periods like a month. Of course, you can set up higher goals for longer periods. For a three-month period, you may set up 25% goals. Another point to note is that large improvements are possible during initial periods. For a programmer, the productivity would be low at the beginning of the working on a new platform, but if the programmer has been working on the same platform for say, two years, the improvement cannot be large. This is because, in two years, the programmer would have mastered the language as well as the tricks of trade in that language. So, while setting goal, consider all such aspects and set realistic goals. The goal should be set such that it is not too easy to achieve, and at the same time, it should not be impossible to reach it.

4. *Benchmarking*: Benchmarking is to compare your performance with that of your peers in your organization and those working in the industry elsewhere. Benchmarking tells you where your performance stands vis-a-vis your peers. Without benchmarking, we may be improving, but it may still fall short of the expected performance. Your performance metrics need to be within 10% of the performance of your peers in the industry. How do you get the data or peer performance? Perhaps, you can request your peers in your organization to share the data on a friendly basis. Most professional organizations compute these metrics and release them in a periodic manner. You can make use of these metrics to benchmark your performance. For data in the industry, the industry associations usually maintain this kind of data and make it available on their websites. Some consultants also maintain this data. Collect data from all such sources to benchmark your performance and then to improve it further.

What is important is to adhere to a defined method in your work, then measure your performance continuously, subject the performance measurements to analysis to draw inferences about your performances, and then studiously improve your performance through careful goal setting.

Index